BO-122

PRINCIPLES OF PLANT SCIENCE

[2 Credits]
Botany : Paper-II
For
First Year B.Sc. Semester-II,
As per New Syllabus CBCS Pattern
June 2019

Dr. K.N. Dhumal
Professor of Botany (Retd.),
Dept. of Botany, SPPU.
Pune – 411 007.

Dr. B.P. Shinde
Principal,
Vidya Pratishthan's Arts, Science
and Commerce College, Baramati.

Dr. H.S. Patil
Asso. Prof. & Head, Dept. of Botany,
Vidya Pratishthan's Arts, Science
and Commerce College, Baramati.

Dr. B.M. Gaykar
Prof. and Head
Dept. of Botany
Ahmednagar College, Ahmednagar

Dr. K.S. Bhosale
Asst. Prof.
Dept. of Botany,
N. Wadia College, Pune 411 004.

Dr. K.N. Gaikwad
Asso. Prof.
Dept. of Botany
K.T.H.M. College, Nashik

N5033

F.Y.B.SC. Principles of Plant Science   **ISBN 978-93-89533-92-7**

First Edition :	**November 2019**
© :	**Authors**

Published By:
NIRALI PRAKASHAN
Abhyudaya Pragati, 1312, Shivaji Nagar
Off J.M. Road, PUNE – 411005
Tel - (020) 25512336/37/39, Fax - (020) 25511379
Email : niralipune@pragationline.com

➢ **DISTRIBUTION CENTRES**

PUNE

Nirali Prakashan : 119, Budhwar Peth, Jogeshwari Mandir Lane, Pune 411002,
(For orders within Pune) Maharashtra, Tel : (020) 2445 2044, Mobile : 9657703145
Email : niralilocal@pragationline.com

Nirali Prakashan : S. No. 28/27, Dhayari, Near Asian College Pune 411041
(For orders outside Pune) Tel : (020) 24690204; Mobile : 9657703143
Email : bookorder@pragationline.com

MUMBAI

Nirali Prakashan : 385, S.V.P. Road, Rasdhara Co-op. Hsg. Society Ltd.,
Girgaum, Mumbai 400004, Maharashtra;
Mobile : 9320129587 Tel : (022) 2385 6339 / 2386 9976,
Fax : (022) 2386 9976
Email : niralimumbai@pragationline.com

➢ **DISTRIBUTION BRANCHES**

JALGAON

Nirali Prakashan : 34, V. V. Golani Market, Navi Peth, Jalgaon 425001,
Maharashtra, Tel : (0257) 222 0395, Mob : 94234 91860;
Email : niralijalgaon@pragationline.com

KOLHAPUR

Nirali Prakashan : New Mahadvar Road, Kedar Plaza, 1st Floor Opp. IDBI Bank,
Kolhapur 416 012, Maharashtra. Mob : 9850046155;
Email : niralikolhapur@pragationline.com

NAGPUR

Nirali Prakashan : Above Maratha Mandir, Shop No. 3, First Floor,
Rani Jhanshi Square, Sitabuldi, Nagpur 440012, Maharashtra
Tel : (0712) 254 7129;
Email : niralinagpur@pragationline.com

DELHI

Nirali Prakashan : 4593/15, Basement, Agarwal Lane, Ansari Road, Daryaganj
Near Times of India Building, New Delhi 110002
Mob : 08505972553, Email : niralidelhi@pragationline.com

BENGALURU

Nirali Prakashan : Maitri Ground Floor, Jaya Apartments, No. 99, 6th Cross,
6th Main, Malleswaram, Bengaluru 560003, Karnataka;
Mob : 9449043034
Email: niralibangalore@pragationline.com
Other Branches : Hyderabad, Chennai

niralipune@pragationline.com | www.pragationline.com

Also find us on   www.facebook.com/niralibooks

Preface ...

The books of Botany published by Nirali Prakashan for F.Y. B.Sc students of Savitribai Phule Pune University are highly popular and are always remained at the top compared to other publications since the annual pattern, semester pattern and now in credit based choice system.

This book of Sem II Paper II. BO-122 : Principles of Plant Science includes "Plant Physiology, Cell Biology and Molecular Biology". It is also written keeping the same tradition. All the topics are written in a highly simplified manner and explained with maximum, well labeled neat diagram. Each chapter is having points to learn, points to remember and exercise. This will help the students for the preparation of final examination. All the sources and reference books referred while writing this book are acknowledged by the authors.

Reading this book is just like the journey through various botanical aspects. The students will feel it as an exciting Academic excursion. We are confident that the topics in this book can be easily understood by our first year students if they have a "Learning attitude". This will help to change you their destiny.

We the team of experienced authors have tried our level best to focus on each and every aspect prescribed by BOS in Botany for CBCS.

From years together we have experienced that "Every student is the Author of his success or failure". Hence understanding the mind of students is of paramount importance, while writing the books of Botany.

Hope the students and their mentors will appreciate our sincere efforts. The suggestions and healthy comments to elevate and enhance the quality, suitability and utility of this book to our beloved students and teachers of Botany will be always very much appreciated. It will inspire all the authors. Your comment as well as criticism is quate valuable for us.

We sincerely thank the expert team of Nirali Prakashan headed by energetic and ever smiling person Mr. Jignesh Furia and Mr. Dineshbhai. All they are the real source of energy while writing this excellent book. They are supporting whole heartly our efforts.

Authors

Message to Student Friends...

❖ *"Remember ! India's Problem is not Economic nor Politics, it is Education only."*

❖ As some one quotes that : **"Indian Education is in ICU"**

❖ *"But it can be improved through the New Mantras of Education :* **Ethics, Honesty and Fairness......"**

❖ Integrity, Dedication and Compassion is the solution.

❖ Our teachers should give a **"Midas touch"** to the students to make them extra ordinary and academically excellent.

❖ Most important in the life of every student is to become employable for which he/she needs to grasp the Mantra **KNOWING, DOING** and **BEING.**

Syllabus ...

BO-122
SEMESTER II: PAPER II
PRINCIPLES OF PLANT SCIENCE

CREDIT-I PLANT PHYSIOLOGY & CELL BIOLOGY 15L(15H)

1. Introduction, definition and scope of plant physiology. 1 L
2. Diffusion – definition, factors affecting diffusion, importance of diffusion in plants, imbibition as a special type of diffusion. 1 L
3. Osmosis – definition, types of solutions (hypotonic, isotonic, hypertonic), endosmosis, exo-osmosis, osmotic pressure, turgor pressure, wall pressure, importance of osmosis in plants. 2 L
4. Plasmolysis – definition, mechanism and significance. 1 L
5. Plant growth and growth regulators – introduction, phases of growth, factors affecting growth, plant growth regulators – introduction, definition and their significance. 2 L
6. Structure of plant cell, differences between prokaryotic and eukaryotic cell. 1 L
7. Plant cell wall – components of primary cell wall, structure and functions. 1 L
8. Plasma membrane- bilayer and fluid mosaic model, components and functions 1 L
9. Ultrastructure and functions of chloroplast, mitochondria and endoplasmic reticulum. 2 L
10. Cell cycle in plants – phases of cell cycle (G_1, M, G_2 and S), importance of cell cycle in plants, divisional stages of mitosis and meiosis. 3 L

CREDIT-II: MOLECULAR BIOLOGY 15L(15 H)

1. Introduction and scope of molecular biology, central dogma of molecular biology. 2 L
2. Structure of DNA- Structure of nitrogen bases, nucleoside, nucleotide, Chargaff's rule, C value paradox. 2 L
3. Watson Crick model of DNA and its characteristic features, types of DNA (A, B and Z DNA). 3 L
4. Packing of DNA into chromosomes, types of chromosomes. 2 L
5. Structure and types of RNA. 3 L
6. DNA replication- Types of replication (conservative, semi-conservative and dispersive), bacterial DNA replication (Initiation, elongation and termination), enzymes involved, leading and lagging strands, Okazaki fragments. 3 L

❖ ❖ ❖

Contents ...

CREDIT - I

1. INTRODUCTION TO PLANT PHYSIOLOGY — 1.1 – 1.9
2. DIFFUSION — 2.1 – 2.6
3. OSMOSIS — 3.1 – 3.10
4. PLASMOLYSIS — 4.1 – 4.4
5. PLANT GROWTH AND GROWTH REGULATORS — 5.1 – 5.20
6. STRUCUTRE OF PLANT CELL — 6.1 – 6.12
7. PLANT CELL WALL — 7.1 – 7.5
8. PLASMA MEMBRNAE — 8.1 – 8.10
9. STRUCTURE AND FUNCTIONS OF CHLOROPLAST, MITOCHONDRIA AND ENDOPLASMIC RETICULUM — 9.1 – 9.14
10. CELL CYCLE IN PLANTS — 10.1 – 10.16

CREDIT - II

11. INTRODUCTION TO MOLECULAR BIOLOGY — 11.1 – 11.4
12. STRUCTURE OF DNA — 12.1 – 12.7
13. WATSON CRICK MODEL OF DNA — 13.1 – 13.8
14. PACKING OF DNA INTO CHROMOSOMES — 14.1 – 14.6
15. STRUCTURE AND TYPES OF RNA — 15.1 – 15.6
16. DNA REPLICATION — 16.1 – 16.11
* REFERENCES — R.1 – R.1
* PATTERN OF UNIVERSITY EXAM QUESTION PAPER — P.1 – P.1
* MODEL QUESTION PAPER — P.2 – P.3
* QUESTION BANK — Q.1 – Q.2

1

CHAPTER

INTRODUCTION TO PLANT PHYSIOLOGY

◆ POINTS TO LEARN ◆

1.1 **Introduction**
1.2 **Definition**
1.3 **Scope and Applications of Plant Physiology**
*　**Points to Remember**
*　**Exercise**

1.1 INTRODUCTION

The food, fodder, fuel, fibres, shelter, clothing, medicines and even oxygen required for human beings, animals and micro-organisms is the outcome of plant physiology. The main physiological processes in plants such as "Photosynthesis" are responsible for the supply of all the above mentioned basic human needs. In spite of the rich diversity of plants and microbes, they share a common master plan of physiological functions, which can be understood by the knowledge of plant physiology. Plant physiology, the most challenging, dynamic and modern discipline of botany, deals with the various metabolic processes and pathways in plants and microbes.

Virtually, plant physiology is the heart of botanical science which is defined as **"the study of various functions performed by plants"** e.g. seed germination, growth and development, absorption of water and minerals, ascent of sap, translocation of solutes, transpiration, photosynthesis, respiration, photorespiration, photoperiodism, vernalization, ripening of fruits, senescence and death of plants.

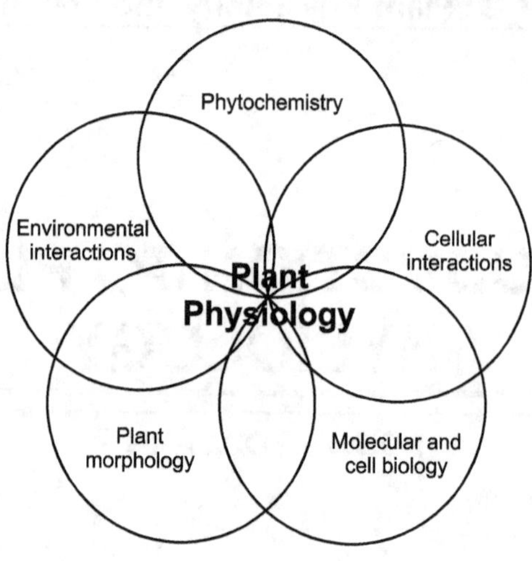

Fig. 1.1

It also covers the study of functions performed by micro-organisms like bacteria and fungi, from the molecular level to the organismic level. In short, plant physiology deals with the study of organisation and operation of all the physiological, biochemical and enzymological processes in plants. It is also concerned with the regulation of development and functioning of plants. The physiological status of a plant gives an idea about its general health, yield and phenotypic nature. Because of the interdisciplinary nature of plant physiology, now-a-days it is also called as experimental botany.

Plant metabolism mainly catabolism and anabolism is studied under plant physiology which is concerned with every aspect of plant life and provides explanation to several questions about the plants e.g. How they use solar energy? How they obtain and distribute water and nutrients? How they grow and develop? How they respond to the environment? How they produce flowers, fruits and seeds? How seeds germinate and form new plants?

The understanding of plant physiology is necessary for the study of all other branches of plant science. The various functional aspects of plants and microbes are studied under plant physiology and all other disciplines depend on it. It is the pivotal branch of botany, which has prime importance in our daily life. The knowledge of physiology of microbes as well as higher plants is most important for their efficient utilization and application in our day-to-day life.

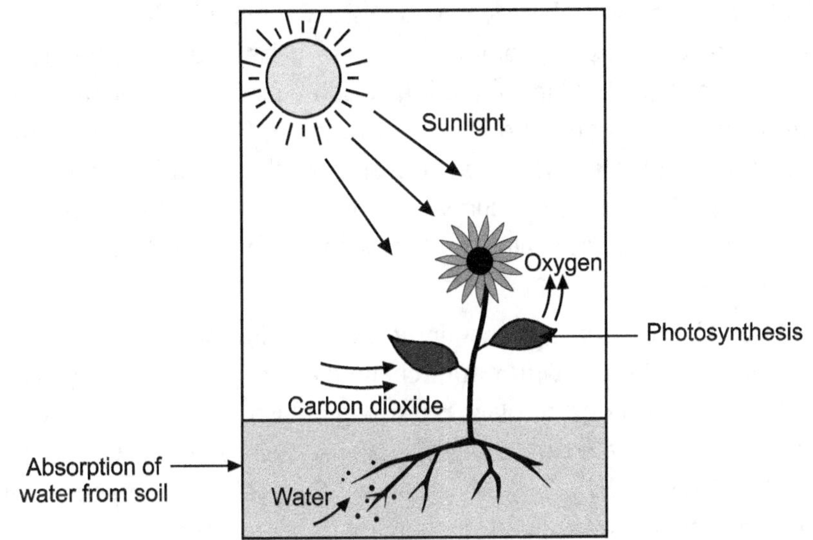

Fig. 1.2: Different functions performed by plant

The growth, development, reproduction, flowering, fruiting and other important processes right from seed germination to death, about every plant is covered in plant physiology. Hence, the knowledge of plant physiology has become a central point in plant science. It is not only a basic science but also its knowledge finds wide applications in every field of commercial botany. The improvement in satisfying the basic needs of human beings and their comfort as well luxury is the outcome of our recent knowledge of plant physiology.

This science essentially finds applications in the fields of biophysics, biotechnology, biochemistry and molecular biology. The knowledge of plant physiology strongly supports other branches of botany such as plant anatomy, cytology, genetics, morphology, ecology, etc. Today, the availability of ultra-modern technology and sophisticated instruments are helping us to upgrade our knowledge about plant physiology.

At the same time, all the above branches become more meaningful, because of the knowledge of plant physiology. The other sciences like plant pathology, agronomy, entomology, genetics and plant breeding, soil physics, soil chemistry, horticulture, sericulture, vegetable science, pomology etc. are also benefited from the study of plant physiology. Plant physiology also helps us to understand the interactions of plants with their environment. Hence, most of the important agro-technologies developed are based on the physiological paradigms. Similarly, plant biotechnology and its approaches are based mainly on the principles and practices of plant physiology.

The above explanations indicate the importance of plant physiology in dealing with future challenges in agriculture and allied fields. The knowledge of plant physiology plays a key role in solving the problems created by climatic changes such as global warming, effect of green house gases, increased CO_2 level, soil, water and air pollution etc. In short, plant physiology has immense importance in safeguarding our environment. Sincere attempts should be made to study plant physiology at each level of learning i.e. from school, college and university.

Today plant physiology has become an important branch for conservation and protection of biodiversity, sustainable development, and improvement in crop yield, safe storage of agricultural produce and improvement in crop productivity under changing climatic conditions.

1.2 DEFINITION

Plant Physiology is the branch of Botany (Plant science) that aims to understand how plants live and function. Actually, it deals with the knowledge of life processes performed by plants throughout their life. This branch of plant science is very helpful to understand all the aspects and manifestation of the plants existing on this planet earth, right from bacteria, fungi, algae, bryophytes, pteriodophytes, gymnosperms and angiosperms. In a simpler way plant physiology is the study of structure and function of plants.

1.3 SCOPE AND APPLICATIONS OF PLANT PHYSIOLOGY

(1) Study of plant physiology has a vibrant scope and importance in different fields such as Agriculture, medicine, food production, textiles, biotechnology, molecular biology and even space science.

(2) It covers the studies right from gene level to organismic level and concerns with genotypic and phenotypic expressions of plants. The fundamental principles and laws of plant physiology are equally important and applicable to all the unicellular and multicellular organisms of either eukaryotic or prokaryotic nature.

The scope of plant physiology is well explained in Fig. 1.3.

(3) Understanding of plant physiology has a great potential to provide jobs and career opportunities to everyone.

(4) Cytogenetics, plant breeding, cell biology, anatomy, morphology, ecology, pathology, pharmacognosy, plant propagation and even plant biotechnology have their basis in plant physiology. All the above branches of fundamental botany are interlinked with plant physiology.

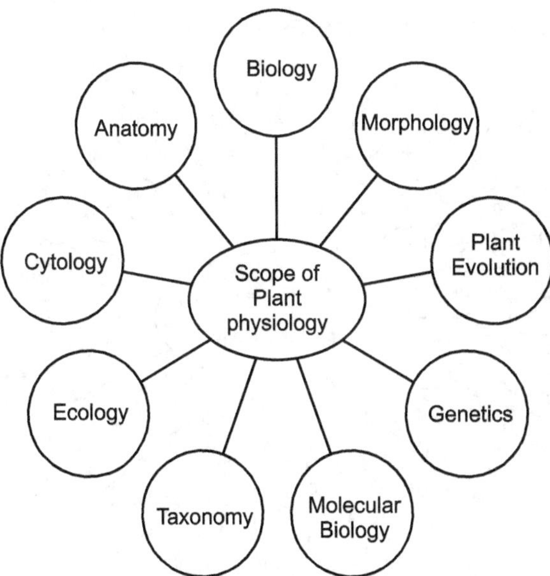

Fig. 1.3: Scope of plant physiology and applications of plant physiology

(5) The molecular and structural changes in plants, various modifications in morphology, resistance of plants to diseases and abiotic stresses like drought, salt, temperature, cold, flood and heavy metals like cobalt, lead, mercury, etc. is well understood and explained on the basis of metabolic principles.

(6) The flowering and fruiting behaviour as well as yielding potential of all plants depend on their physiological specificity or physiological adaptations.

(7) The physiological status of plants indicate their yield, productivity, survival potential as well as adaptations to biotic as well as abiotic stress conditions.

(8) The knowledge of plant physiology has very vast scope in every field and its applications play important roles in our day to day life (Fig 1.4).

(9) Plant physiology has great scope in many applied sciences like biochemistry, biophysics, phytochemistry, biostatistics, agriculture, computer and electronic science and even in space science. This subject has very strong and well established links with above mentioned modern sciences (Fig 1.4).

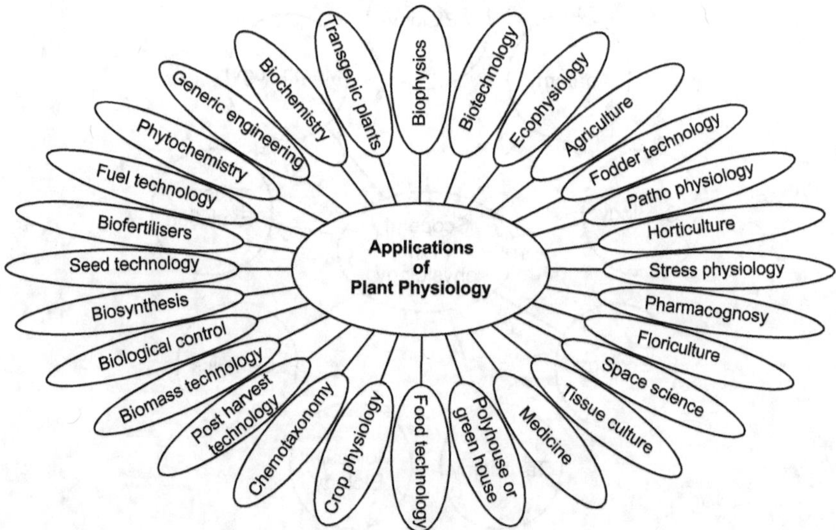

Fig. 1.4: Applications of plant physiology in different fields of science

(10) At present, plant physiology has tremendous scope in development of high tech agriculture, floriculture, horticulture, seed technology, post harvest technology, polyhouse or greenhouse technology, tissue culture, biomass technology, food technology, fuel technology, biological control, genetic engineering and production of transgenic plants, production of drought, salt, flood, temperature and radiation resistant plants (stress tolerant plants). The important applications of plant physiology are summarised in Fig. 1.4.

(11) The most amazing world of plant science is well explored, explained and wisely exploited for the benefit of man through the knowledge of plant physiology.

(12) Plant physiology is mostly an application oriented subject with great potential for national development. It is difficult to point out the life processes and life improvement programmes in which plant physiology is not involved.

(13) Human beings require food, shelter, clothing and medicines for health care, and many other things for improving the quality of life, which is mainly achieved through the proper application of knowledge or principles of plant physiology.

POINTS TO REMEMBER

- Plant physiology is the science to understand basic structure and functions of all the plants existing on the planet earth.

- Plant physiology is defined as the branch of botany dealing with structure and function of plants. Its the journey of botanists to explore that how plants live and functions.

- This branch is very well connected to other branches and aspects of botany like morphology, anatomy, cytology, genetics etc.

- It has great scope in every field of science, e.g. Agriculture, medicine, food production, textiles, biotechnology, molecular biology and even space science.

- This branch has a great potential to provide jobs to everyone.

- The food, fodder, fuel, fibres, shelter, clothing, medicines and even oxygen required for human beings, animals and micro-organisms is the result of plant physiology.

- Plant physiology is the heart of botanical science which is defined as the study of various functions performed by plants e.g. seed germination, growth and development, absorption of water and minerals, ascent of sap, translocation of solutes, transpiration, photosynthesis, respiration, photorespiration, photoperiodism, vernalization, ripening of fruits, senescence and death of plants.

- The other sciences like plant pathology, agronomy, entomology, genetics and plant breeding, soil physics, soil chemistry, horticulture, sericulture, vegetable science, pomology etc. benefit from the study of plant physiology.

- It also deals with the fundamentals of plant's life i.e. seed germination, growth, flowering, seed formation, storage of food material, productivity etc.

- Plant physiology helps us to understand the interactions of plants with their environment.

- The knowledge of plant physiology plays a key role in solving the problems created by climatic changes such as global warming, effect of green house gases, increased CO_2 level, soil, water and air pollution etc.

- Today, plant physiology has become an important branch for conservation and protection of biodiversity, sustainable development, and improvement in crop yield, safe storage of agricultural produce and improvement in crop productivity under changing climatic conditions.

EXERCISE

1. Give definition of plant physiology and explain how it is linked to other branches of plant science.

2. Explain the scope of plant physiology in different fields of day to day life of human beings.

3. Comment on :

    (i) Scope of plant physiology in post harvest technology.

    (ii) Hightech agriculture

    (iii) Horticulture

(iv) Biomass technology

(v) Food technology

(vi) Production of stress tolerant crops.

4. Justify, "Plant physiology in real sense is experimental biology".

5. Define plant physiology. Explain why to study plant physiology?

6. Describe the scope and importance of plant physiology.

7. Give the applications of plant physiology in various fields.

2

CHAPTER

DIFFUSION

♦ POINTS TO LEARN ♦

2.1 Introduction
2.2 Definition
2.3 Factors Affecting Diffusion
2.4 Importance of Diffusion in Plants
2.5 Imbibition
    2.5.1 Concept
    2.5.2 Definition
    2.5.3 Imbibition as a Special Type of Diffusion
*   **Points to Remember**
*   **Exercise**

2.1 INTRODUCTION

For the survival of all plants, water is essential which plays predominant role in every plant process. Both accute shortage (drought) and excess (flooding) of water availability is culminating the plants in to death. Hence, balance of water has the life of plants in top most priority. Large number of researchers are focussing this problem throughout the world. To understand the plant water relationship understanding of basic processes involved in it e.g. Osmosis, Diffusion, Imbibition must be understood in depth by plant scientists.

Almost all of us have experienced the phenomenon of diffusion in one way or the other. When we put sugar in a liquid, such as coffee or tea, the sugar molecules diffuse throughout the liquid giving it a uniform sweet taste. If a bottle of essential oil is opened, molecules of the perfume diffuse through the molecules of air. Similarly if a bottle of acetic acid, chloroform, formaldehyde, etc. remain open in a laboratory there is a diffusion phenomenon. In a chemistry laboratory, hydrogen sulphide gas apparatus is commonly found through which H_2S gas diffuses into the surrounding atmosphere and produces the foul smell of rotten egg. This is due to the fact that the molecules of the gas have a tendency to move from the regions of higher concentration to the regions of lower concentration. This goes on until the equilibrium is established throughout the available space.

2.2 DEFINITION

Literally speaking, to diffuse means to *disperse* or *to spread* or *to flow out* or *to extend in all directions*. As mentioned above, like the gases, for example, if solute is placed (i.e. sugar) in a solvent (i.e. water), solute dissolves completely. The particles of solute (sugar) distribute (i.e. diffuse) uniformly. This is called as diffusion. Thus diffusion is defined as, "*The movement of particles or molecules from a region of higher concentration to a region of lower concentration; Or It is the movement of molecules from a region of high partial pressure to the region of low partial pressure as a result of their inherent kinetic energy.*" It also can be defined as the "*tendency of the molecules of liquid, solid or gas to distribute themselves evenly throughout the available space.*" In a simplest way it is the phenomenon of movement of particles of matter due to their own kinetic energy.

In a nutshell "*diffusion is the movement of molecules from high concentration to the region of their low concentration due to their inherent kinetic energy, till they get distributed uniformly until a dynamic equilibrium is established.*" (Fig. 2.1)

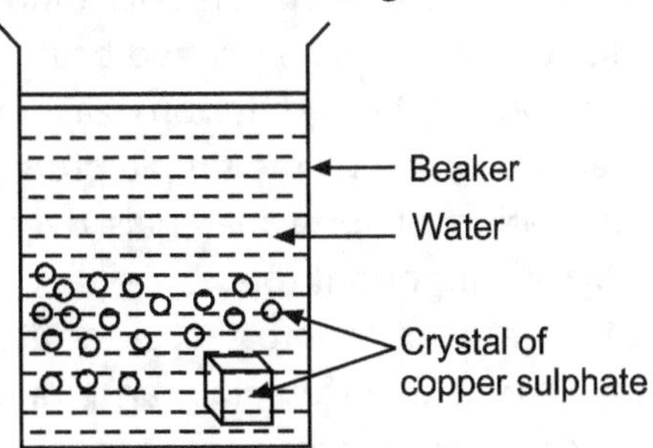

Fig. 2.1: Process of Diffusion

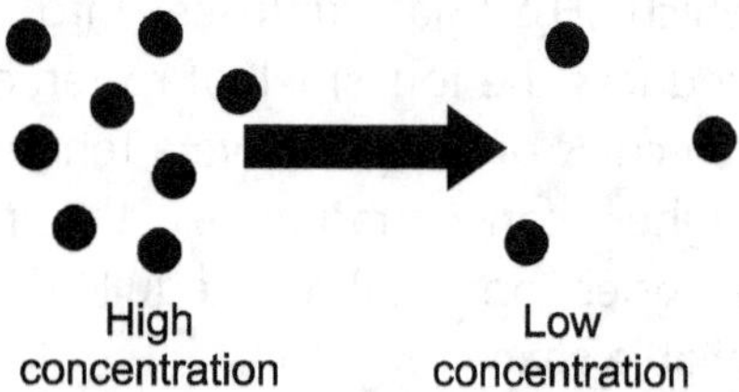

Fig. 2.2: Diffusion of a solid in water

2.3 FACTORS AFFECTING DIFFUSION

Following are the important factors affecting diffusion:

- The rate of diffusion increases with the increase in temperature.
- The diffusion pressure gradient affects the process of diffusion. If the gradient increases, the rate of diffusion also increases. In other words, the rate of diffusion increases with the steepness of diffusion pressure gradient.
- Diffusion depends on the concentration or the number of diffusing particles. The greater the concentration of diffusing particles in a system, greater will be the diffusion pressure and diffusion itself.
- Chemical nature of diffusing substances: whether the diffusing substance is acidic, basic, electrolyte or a non-electrolyte etc. will play an important role in diffusion.

2.4 IMPORTANCE OF DIFFUSION IN PLANTS

In plant physiology, phenomenon of diffusion has great significance in the life of a plant as mentioned below.

- In passive salt absorption, diffusion plays a significant role where ions are absorbed by the simple process of diffusion.
- In the translocation of food materials over short distances, diffusion plays an important role.
- Diffusion of water vapour into the outer atmosphere occurs through open stomata in stomatal transpiration.
- Exchange of gases during the process of photosynthesis and respiration occur through the same stoma, simultaneously, through the process of independent diffusion.

2.5 IMBIBITION

2.5.1 Concept

Like diffusion and osmosis, imbibition is also actively involved in water uptake under certain circumstances. The swelling of dry seeds, when they are put in water is the best example of imbibition of water. Similarly imbibition of wood when kept in water is also another example of this phenomenon. The wooden doors and windows get jammed during rainy season because of the imbibition of water in them.

Imbibition is one of the physical process available to the plant, for the uptake of water. If a small quantity of seeds is placed in water, they absorb water. They swell and there is an increase in the volume of the seeds. This is called as imbibition. Thus "imbibition is a phenomenon of taking in of water and consequent swelling of colloidal materials."

2.5.2 Definition

Absorption of water by colloidal materials resulting into the swelling of colloidal materials is known as imbibition. The solvents which are usually imbibed into the materials have affinity or attraction for them (hydrophilic colloids). These forces of attraction are usually chemical or electrostatic forces. The substances are generally hydrophilic. The movement of water is along the diffusion gradient. Imbibition involves both absorption as well as adsorption of water or solvents in colloidal substances like proteins, agar, cellulose, gelatin, starch etc. The substance which imbibes water is called as *imbibant.*

Imbibition process can be defined as given below:

"The uptake or absorption of liquids by the solid particles of a substance without forming a solution is called Imbibition".

2.5.3 Imbibition as a Special Type of Diffusion

In imbibition there is both absorption as well as adsorption. Water enters into the cell, thereby increasing the volume. This is called as absorption. Some water molecules are held over the surface. This is known as adsorption. The molecules which are tightly held on the surface of the imbibitant lose their kinetic energy due to immobility. Thus, when seeds are soaked in water there is warming of water. This is known as "heat of wetting."

As already mentioned, the "absorption of water by hydrophilic colloids is called as imbibition." The movement of water is along the diffusion gradient. "Imbibition is the capacity of a gel or any other colloidal material to take up relatively large quantities of water and swell."

E.g. Swelling of doors and wood work during the rainy season, absorption of water by cell wall, swelling of seed coats, starch, cellulose, agar, gelatin, etc. If agar is added to water, within a few minutes, it absorbs sufficiently large quantities of the water. The phenomenon of imbibition increases volume of imbibant due to which pressure is created is called as *imbibition pressure*. It is the potential maximum pressure which an imbibant will develop if it is submerged in water. All this clearly indicates that imbibition is a special type of diffusion in plants. During imbibition the molecules diffuse as a result of their concentration gradient till they get distributed uniformly.

Table 2.1: Difference between Diffusion and Imbibition

Diffusion	Imbibition
1. This is simple dispersion phenomenon of different states of matter.	1. This is a physical phenomenon of hydrophilic matters.
2. Diffusion particles have different diffusion pressure.	2. Imbibant develops a imbibition pressure.
3. Diffusion takes place from the region of higher concentration to the region of lower concentration.	3. Imbibition occurs in substances possessing hydrophillic colloides (insoluble organic matters)
4. It occurs in solids, gases, liquids, etc.	4. It occurs in organic solids like wood, seeds etc.
5. Temperature plays a prominent role in this.	5. Humidity plays an important role in it.
6. e.g. Exchange of gases during photosynthesis and respiration.	6. e.g. Swelling of wood, germinating seeds, etc.

POINTS TO REMEMBER

- Diffusion is the movement of molecules from high concentration to the region of their low concentration due to their inherent kinetic energy till they get uniformly distributed until a dynamic equilibrium is established.

- The best example is opening of a bottle of essential oil (Attar) in the air. The molecules of perfume diffuse through the molecules of air.
- Important factors affecting diffusion are (i) temperature, (ii) diffusion pressure gradient, (iii) concentration of diffusing molecules, (iv) chemical nature of diffusing substance.
- The process of diffusion is very important in plant's life e.g. absorption of salt, translocation of food materials stomatal transpiration, exchange of gases in photosynthesis and respiration takes place due to diffusion.
- Absorption of water by colloidal materials resulting into the swelling of colloidal material is known as Imbibition.
- It is the process helping in uptake of water.
- The uptake or absorption of liquids by the solid particles of substance without forming a solution is called Imbibition.
- Imbibition is the absorption of water by hydrophilic colloids, e.g. swelling of doors and wood works during the rainy season, absorption of water by cell wall.
- The rate of diffusion is proportional to the kinetic energy of the molecules, their size, the density of the medium through which they pass and the gradient of concentration over which they diffuse.

EXERCISE

1. Define and discuss the process of diffusion in plants.
2. Comment on: Factors affecting diffusion.
3. Give an account of importance of diffusion in plants.
4. Distinguish between diffusion and imbibition.
5. Comment on imbibition.
6. Define and discuss the phenomenon of imbibition.
7. Describe the mechanism of diffusion and explain the factors affecting it.
8. Justify: Imbibition is a special type of diffusion.

3

CHAPTER

OSMOSIS

♦ POINTS TO LEARN ♦

3.1 **Introduction**
3.2 **Definition**
3.3 **Types of Solutions (hypotonic, isotonc, hypertonic)**
3.4 **Osmotic Pressure**
3.5 **Types of Osmosis**
3.6 **Turgor Pressure**
3.7 **Wall Pressure**
3.8 **Importance of Osmosis in Plants**
* **Points to Remember**
* **Exercise**

3.1 INTRODUCTION

Osmosis is a fundamental physiological phenomenon in plants. It is involved in water absorption, translocation of nutrients in cells, etc. It is a special type of diffusion. Osmosis involves the movement of water through a differentially permeable membrane from an area where it is in high concentration to an area where it is in low concentration.

Osmosis is also called as osmotic diffusion. It is a phenomenon of diffusion through a semi-permeable membrane.

3.2 DEFINITION

It is defined as, *"When two solutions of different concentrations are separated by means of a semi-permeable membrane, the diffusion of water or solvent takes place from the solution of lower concentration to the solution of higher concentration, until a state of dynamic equilibrium is attained"*.

Actually, it is observed that diffusion of the solvent takes place both ways across the membrane but the diffusion of the solvent is

(3.1)

more from the solution of lesser concentration to that of the higher concentration.

Thus in osmosis, semi-permeable membranes, concentrations of solutions and concentration gradients play an important role.

3.3 TYPES OF SOLUTIONS

A solution is a homogeneous mixture of two substances: a solute and a solvent.

(a) Solute: substance being dissolved; present in lesser amount.

(b) Solvent: substance doing the dissolving; present in larger amount.

Solutes and solvents may be of any form of matter: solid, liquid or gas.

Different types of solutions have different chemical properties. The concentration of a solute in a solution is a measure of how much of that solute is dissolved in the solvent. On the basis of the concentration of the solute in a solution, solutions are categorised as hypotonic, hypertonic, and isotonic.

Hypotonic:

The solution which contains more solute than solvent [example: a lot of salt (solute) dissolved in water (solvent)].

Hypertonic:

The solution which contains more solvent than solute [example: In purified water there is almost no solute dissolved in the solvent (water)].

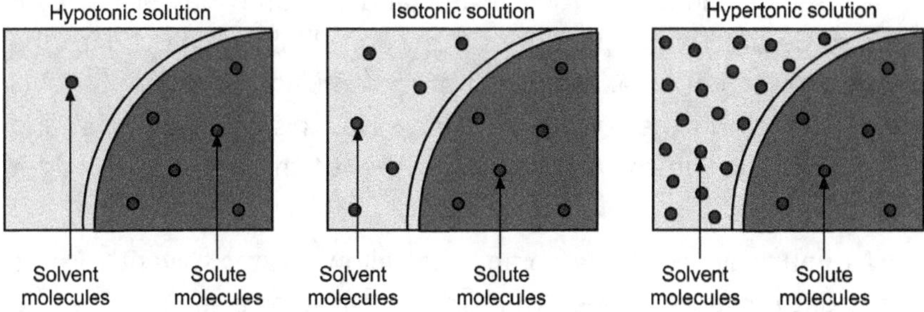

Fig. 3.1: Types of Solutions

Isotonic:

The solution in which the solute and solvent are equally distributed--a cell normally wants to remain in an isotonic solution, where the concentration of the liquid inside equals the concentration of the liquid outside it.

Mechanism of Osmosis:

In osmosis, the movement (diffusion) of solvent takes place from a solution having high concentration of solute to a solution having low concentration of solute through a semipermeable membrane or a selectively permeable membrane like the plasma membrane, tonoplast, etc. This movement of solvent is always along the concentration gradient. In other words, osmosis is the diffusion of water from weak solution into the strong solution but through semipermeable membrane.

Osmotic pressure is responsible for the movement of solvent through the semi-permeable membrane; which is proportional to the concentration of solute molecule in a given amount of solvent.

Demonstration of Osmosis

Take a U-shaped tube and in the middle fix a semi-permeable membrane. Fill the right limb of the U tube (A) with 10 ml of water and left limb of U tube (B) with 10 ml of 10 per cent sugar solution. The water molecules will only diffuse freely across the semi-permeable membrane from tube A into B. The level of solution in arm B will increase. (Fig. 3.2 A and B)

Thus, right limb has 100% water and left limb has 90% water. After some time water molecules from right limb will move into the solution in the left limb. As a result of the diffusion of water, the level of the solution in the left limb will increase. This will cause lowering of concentration in the left limb and level of the water in the right limb.

As you may recall, membrane that allows water solvent to pass freely, but prevents passage of solutes is called as a semi-permeable membrane and the diffusion of water or any solvent (not the solution) through a semi-permeable membrane is called as osmosis.

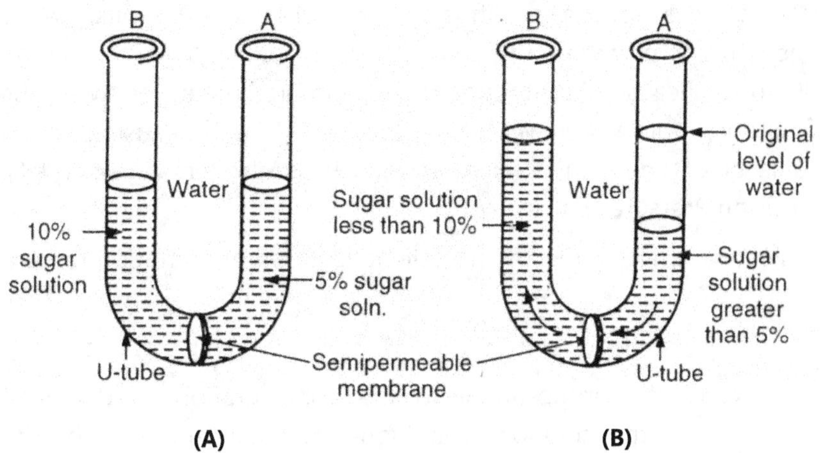

Fig. 3.2: Physical demonstration of osmosis by using pure solvent

Thus, from the above description, it can be concluded that in general, three conditions are necessary for osmosis:

- A strong solution,

- Water or a weak solution, and

- A semi-permeable membrane between them.

Osmosis is thus diffusion of water (solvent) into the solution or from a weak solution into a strong solution. Truly semi-permeable membranes are rare and hence some diffusion of solute also takes place. These membranes are hence called as selectively or differentially permeable membranes.

3.4 OSMOTIC PRESSURE

In the above U-tube experiment (Fig. 3.2), as a result of the separation of solution from it's solvent or the two solutions by the semi-permeable membrane, a pressure is developed in the solution due to presence of dissolved solutes in it. This is called as *Osmotic pressure*. Because of O.P. there is rise in the level of the liquid in the "B" limb of the tube.

The development of O.P. is the result of the difference in the diffusion pressure of water molecules on the two sides of the semi-permeable membrane. The concentration of water in the "A" limb is

90% (U-tube) so say diffusion pressure is 90 units, while concentration of water in the left "B" limb is 100% so D. P. would be 100 units. Because of difference in diffusion pressure in the two limbs of U-tube, more water will diffuse into the "B" limb, causing rise in the level of liquid in it. Here thus, O.P. is also known as D.P.D. (Diffusion Pressure Deficit).

Following are some of the salient features of osmotic pressure (O.P.):

- O.P. of a particular solution is always higher than it's pure solvent.

- It is directly proportional to the concentration of dissolved solutes in the solution. The higher the concentration of the solution, higher will be it's O. P.

- O.P. is measured in terms of atmospheres.

- It does not increase by the addition of insoluble solutes in the solution.

- O.P. is usually denoted by the symbol Ψ ('Psi') and is measured by Pfeffer's osmometer.

3.5 TYPES OF OSMOSIS

The osmotic movement of a solvent in a cell depends upon the external conditions of the cell. If the cell is placed in a hypotonic solution, water from outside will enter into the cell by a process known as **endosmosis**. (Fig. 3.3 A)

If the same cell is placed in a hypertonic solution, the water from cell will come out into the external solution in a process known as **exosmosis** (plasmolysis). (Fig. 3.3 B)

But if a cell is kept in an isotonic solution there will be no entry or exit of water.

The process of osmosis is closely related to O.P., T.P. and D.P.D.

The absorption of water in almost all plants depends on osmosis, which is known as osmotic absorption of water. It is also known as active absorption of water.

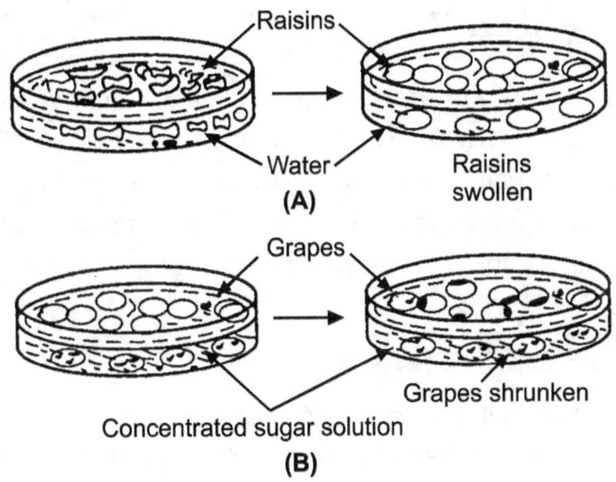

Fig. 3.3: (A) Endosmosis (B) Exosmosis

When dry grapes are kept in water, they absorb water from outside. They become swollen. This is called as phenomenon of endosmosis. Thus, raisins (i.e. dry grapes) swell due to the phenomenon of endosmosis. On the other hand, if fresh grapes are kept in a sugar solution of higher concentration, grapes shrink. This phenomenon is known as exosmosis.

Table 3.1: Difference between Exosmosis and Endosmosis

Exosmosis	Endosmosis
1. When water comes out of the cell sap it is called as exosmosis.	1. When water enters inside the cell sap it is known as endosmosis.
2. This phenomenon occurs when cells are placed in hypertonic solutions.	2. This phenomenon occurs when cells are placed in hypotonic solution.
3. This occurs due to osmotic pressure lower than that of cell sap.	3. This occurs due to osmotic pressure higher than that of the cell sap.
4. This phenomenon causes flaccidity of cells.	4. This phenomenon increases turgidity of cells.
5. It causes plasmolysis of cells.	5. It causes deplasmolysis of cell.

3.6 TURGOR PRESSURE (T. P.)

It is defined as *the pressure developed on the cell wall due to turgidity in the cell.* When a cell is kept in a hypotonic solution, water from outside enters into the cell due to which the cell become turgid and develops a pressure called as *turgor pressure* (T.P.). When equilibrium is established entry of water stops, because in that condition O.P. = T.P.

3.7 WALL PRESSURE (W.P)

When a plant cell is kept in pure water, the external water enters into the cell due to difference in the concentration gradient. As a result of this, the flacid cell becomes turgid. The cell wall which is elastic develops a pressure which is known as wall pressure (W.P.). The wall pressure (W.P.) presses the membrane in the opposite direction. Generally the T.P. is equal to W.P. (T.P. = W.P.)

3.8 IMPORTANCE OF OSMOSIS IN PLANTS

- The absorption of water by the root hairs from soil and the movement of water from one living cell to another within the plants are partly brought about by osmosis. Thus, osmosis plays an important role in plant water relations.

- Osmosis is also mainly responsible for the movement of water from the non-living xylem elements into living cells.

- The turgor developed in the plants is mainly due to osmosis, which makes growth possible in plants. It is also responsible for keeping young stems erect and leaves in the extended form to receive sunlight during photosynthesis.

- Osmosis is responsible for the developing force in growing tissues, especially in young roots, which help them to grow in hard soil layers. This force also helps young seedlings to break the soil and emerge out of it.

- The growth and turgor movements e.g. opening and closing of flowers, sleep movement in leaves are due to osmosis during the process of seed germination

- Osmosis controls the distribution of water in the different parts of the plants after it has been absorbed from the soil.
- The osmotic concentration of cell plays an important role in determining the resistance of plants to drought and frost.

Table 3.2: Difference between Diffusion and Osmosis

Diffusion	Osmosis
1. This is a simple dispersion phenomenon of different states of matter.	1. This is special type of diffusion phenomenon.
2. Diffusion particles have different diffusion pressure.	2. Osmosis is caused by osmotic pressure.
3. Diffusion takes place from the region of higher concentration to the region of lower concentration.	3. Osmosis takes place from the area of high concentration separated by differentially permeable membrane to the area of low concentration.
4. It occurs in solids, gases, liquids, etc.	4. It occurs only in liquid systems that are separated by a semi-permeable membrane
5. Diffusion acts as a one way phenomenon.	5. Osmosis is a two way phenomenon.
6. Temperature plays a prominent role in this.	6. Concentration of solution plays an important role in this.
7. e.g. Exchange of gases during photosynthesis and respiration.	7. e.g. Water absorption by plant cells, opening and closing of stomata etc.

POINTS TO REMEMBER

- Osmosis is defined as, when two solutions of different concentrations are separated by a semi-permeable membrane the diffusion of water takes place from the solution of lower concentration to the solution of higher concentration until a state of dynamic equilibrium is established.

- In osmosis the movement of solvent is always along the concentration gradient.
- Osmotic pressure (OP) is responsible for the movement of solvent through semi-permeable membrane which is proportional to the concentration of solute molecules in a given amount of solvent.
- OP of a particular solution is always higher than its pure solvent. OP is also known as diffusion pressure deficit (DPD).
- Wall pressure (WP) is the pressure developed on the cells due to the entry of external water in to the cell because of difference in the concentration gradient.
- A solution which contains more solute than solvent is known as an hypotonic solution.
- A solution which contains more solvent than solute is known as hypertonic solution.
- A solution in which the solute and solvent are equally distributed is known as an isotonic solution.
- Pressure developed in the solution due to presence of dissolved solutes in it is called as Osmotic pressure.
- If the cell is placed in a hypotonic solution, water from outside will enter into the cell by a process known as endosmosis.
- If a cell is placed in a hypertonic solution, the water from cell will come out into the external solution in a process known as exosmosis (Plasmolysis).
- Turgor Pressure (T. P.) is defined as the pressure developed on the cell wall due to turgidity in the cell.
- OP is the pressure developed in a strong solution, which draws water from weak solution during osmosis.

EXERCISE

1. Define osmosis and explain its mechanism.
2. Write notes on:
   (a) Types of solutions
   (b) Endo and exosmosis

    (c) TP, WP

    (d) Imbibition

3. Explain the role of osmosis in plants.

4. Distinguish between:

    (a) Osmosis and diffusion

    (b) Endo-osmosis and exo-osmosis

    (c) Diffusion and imbibition

    (d) Hypotonic and Hypertonic solution

5. Explain the process of osmosis in plant with suitable experiment.

6. Define osmotic pressure and give its important features.

7. Define Turgor pressure.

8. Explain what is wall pressure.

9. Give an account of different types of solutions.

10. Comment on:

    (a) Hypotonic

    (b) Hypertonic

    (c) Isotonic solutions.

❖ ❖ ❖

PLASMOLYSIS

◆ POINTS TO LEARN ◆

4.1 Introduction
4.2 Definition and Mechanism
4.3 Significance of Plasmolysis in Plants
*** Points to Remember**
*** Exercise**

4.1 INTRODUCTION

In a normal plant cell, protoplasm is tightly pressed against the cell wall. If this normal plant cell is placed in a concentrated solution (i.e. hypertonic solution), water comes out from the cell sap into the outer solution due to exo-osmosis because of the permeability of the cell wall. Protoplasm shrinks or contracts from the cell wall. This is called as *incipient plasmolysis* or *plasmolysis*.

4.2 DEFINITION AND MECHANISM

If the outer solution is more concentrated as compared to the cell sap i.e. it is hypertonic, the protoplasm contracts or shrinks more and more due to exosmosis. The protoplasm due to more shrinkage gets separated from the plant cell wall. It becomes spherical in form. This phenomenon is called as *plasmolysis* (Fig. 4.1).

Plasmolysis (plasma = cytoplasm, lysis = breakdown)

The tissue or cell is said to be plasmolysed. In such a plasmolysed cell, outer space is occupied by hypertonic solution (Fig. 4.1 A to D).

If a plasmolysed cell is placed in water, the phenomenon of endosmosis takes place. Water enters into the cell sap, the cell becomes turgid and protoplasm assumes it's original shape. Such a phenomenon is called as ***deplasmolysis***.

Thus, plasmolysis takes place in three stages - incipient plasmolysis, evident plasmolysis and final plasmolysis (Fig. 4.1 C).

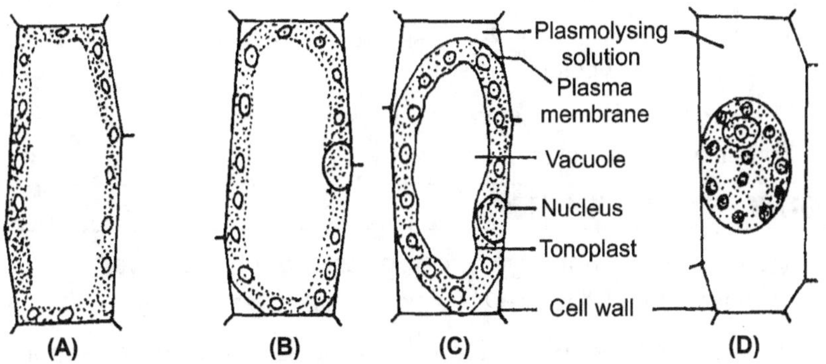

**Fig. 4.1: Plasmolysis: A, turgid cell in water;
B – D, cell after immersion in a strong solution of sugar,
showing successive stages in plasmolysis**

In *incipient plasmolysis*, if a normal cell is kept in a hypertonic solutions there is continuous movement of water from the cell to outside and there is extreme shrinkage of plasma membrane called as *incipient plasmolysis* (Fig. 4.1 B). The cell under *evident plasmolysis*, loses water continuously and shrinks further in volume. The plasma membrane gets torn completely from the cell wall and goes on contracting. This is called as evident plasmolysis (Fig. 4.1 C).

In the last stage of plasmolysis, due to continuous *exosmosis*, the shrinkage of the cell and cytoplasm reaches the maximum limit. The further shrinkage in volume is not possible. Cytoplasm will be completely free from the cell wall and remains in the centre of the cell. This is called as *final plasmolysis*. A cell which has undergone plasmolysis is called *plasmolysed cell* (Fig. 4.1 D).

4.3 SIGNIFICANCE OF PLASMOLYSIS IN PLANTS

The phenomenon of plasmolysis is an important process in plant life. This process has the following significance.

- It proves that cytoplasm acts as a semi-permeable membrane between external solution and vacuole.

- It is evident only in living cells hence the process is useful for deciding whether the cell is living or dead.

- It is used for determining approximate osmotic pressure of the cell by identifying incipient plasmolysis stage.

- The rate of spontaneous deplasmolysis is a measure of the permeability of cell to the solute.
- The process of deplasmolysis is useful to regain original shape and size by the wilted plant organs.

POINTS TO REMEMBER

- Breakdown or shrinkage of cytoplasm from the cell wall is known as plasmolysis (Plasma-cytoplasm, lysis – breakdown).
- Plasmolysis takes place due to exosmosis when the cell is placed in hypertonic solution.
- The tissue or cell in which protoplasm is completely shrinked is known as plasmolysed.
- When plasmolysed cell is kept in water then endosmosis takes place. Water enters in the cell sap, the cell becomes turgid and protoplasm assumes it's original position. This phenomenon is known as deplasmolysis.
- Plasmolysis takes place in three stages:
  (i)　Incipient plasmolysis,
  (ii)　Evident plasmolysis,
  (iii)　Final plasmolysis.
- It takes place only in living cell.
- It is useful to determine O.P.
- Plasmolysis is very important in plant's life.
- It is used to determine O.P. of a cell.
- Deplasmolysis is useful to plants to regain the original shape and size.
- It helps to understand whether the cell is living or dead.

EXERCISE

1. Define plasmolysis.
2. Explain the concept of plasmolysis.
3. Explain how plasmolysis takes place in plant cell.

4. What is incipient plasmolysis.

5. Explain what is mean by plasmolysed and deplasmolysed cell.

6. What is evident plasmolysis?

7. Explain with suitable diagram the process of plasmolysis in plant cell.

8. Explain the significance of plasmolysis in plants.

9. Distinguish between:

    (a) Osmosis and plasmolysis.

    (b) Diffusion and plasmolysis.

    (c) Imbibition and plasmolysis.

5
CHAPTER

PLANT GROWTH AND GROWTH REGULATORS

♦ POINTS TO LEARN ♦

5.1 **Introduction**

    5.1.1 **Definition**

5.2 **Phases of Growth**

5.3 **Factors Affecting Growth**

    5.3.1 **External Factors**

    5.3.2 **Internal Factors**

5.4 **Plant Growth Regulators (PGRs)**

    5.4.1 **Introduction**

    5.4.2 **Definition**

5.5 **Practical Applications of Auxins (IAA)**

5.6 **Practical Applications of Cytokinins**

5.7 **Practical Applications of Gibberellins (GA$_3$)**

5.8 **Practical Applications of Ethylene**

5.9 **Practical Applications of Abscisic Acid (ABA)**

* **Points to Remember**

* **Exercise**

5.1 INTRODUCTION

Growth is an essential character of all living organisms. Development of an entire organism from a single cell or in the higher plants development of mature plant from a tiny seed is said to be growth. There is an increase in size, shape, weight, height, length and biomass due to growth. It is a permanent and irreversible change in an organism. Growth occurs due to the cell enlargement and cell division. It involves various physiological and bio-chemical processes. It needs favourable external and internal environmental conditions.

5.1.1 Definition

Growth can be defined as *"Permanent and irreversible increase in an organism"*. Or *"Growth is a vital process which brings about a permanent change in any organism or its part with respect to its size, weight and volume."*

Growth in plants and animals differ in the following points.

1.   Growth in plants is due to a special tissue known as 'meristem' which is not present in animals.

2.   In plants, growth is localised in meristems present in the buds, at nodal regions and in the vascular bundles. In animals, growth is not localised and is uniform.

3.   In perennial plants, growth is a continuous process responsible for unlimited growth. In animals, growth occurs at a particular period in the life and it is limited.

5.2 PHASES OF GROWTH

The term growth usually refers to quantitative increase in the plant body. The increase in height, fresh and dry weight etc. are the different parameters of growth in plants. The increase in growth parameters is generally irreversible. The growth in plants is restricted to certain regions only, which are known as **growth regions**. In these growth regions, food is constantly accumulated and assimilated at the cost of reserve material, some of which is also oxidised during respiration and release energy in the form of ATP. Growth takes place using this energy. In growth and development, **source-sink** relation is very important. The growing regions show constant increase in dry weight.

The qualitative changes taking place in plant body such as germination of seeds, formation of flowers and fruits, falling of leaves etc. is known as **development** in plants. Growth and development, in fact, cannot be separated into two distinct phenomena occurring at different times. On the other hand, growth follows development in quick succession. When the meristem changes from vegetative to reproductive stage, the change is termed as development. In real sense, both growth and development are integrated and interrelated processes. The study of development from a morphological point of view is known as **morphogenesis**.

Growth is one aspect of development of an organism. It begins with a single cell. In plants, meristematic cells have to pass through three phases of growth.

(i) Cell formation or exponential phase.

(ii) Cell elongation or linear phase.

(iii) Cell maturation (differentiation) or senescence phase.

Each phase is characterised by some physical and physiological changes in the plant. The time period required to complete each phase is specific.

(i) Cell formation phase: It is the first phase of growth, in which after cell division, new cells are formed. These cells are small in size, have an active nucleus and cytoplasm. They are ready to absorb water, food material and have a high metabolic rate. Cell walls are thin and cells contain a large quantity of solutes.

(ii) Cell elongation phase: After accumulation of a large quantity of solutes, the cells absorb water by osmotic process, which result in increasing turgidity of the cells and expansion of thin and elastic cell walls. This results in increase in the volume of the cells. More cellulose molecules get deposited on the cell walls due to which more water is absorbed and cells become turgid. Turgidity increases due to appearance of a large central vacuole in each cell. It leads to increase in length and size of the cell.

(iii) Cell maturation phase: In this phase, metabolic activities of the cells increase and the cells accumulate food material. Secondary cell walls are laid down and the cells mature and get differentiated into tissues. There is no change in the size of the cells. Many biochemical changes occur in the cells and cells enter the senescence phase. Growth phases can be clearly seen in a germinating seed. (Fig. 5.1)

Growth is slow at the beginning (at cell formation phase), becomes rapid later (at cell elongation phase) and again slows down at the end (at cell maturation phase) and attains a steady stage. If a graph is plotted by considering growth rate against time for any plant, it results in a 'S'-shaped curve known as *the 'Sigmoid curve'*.

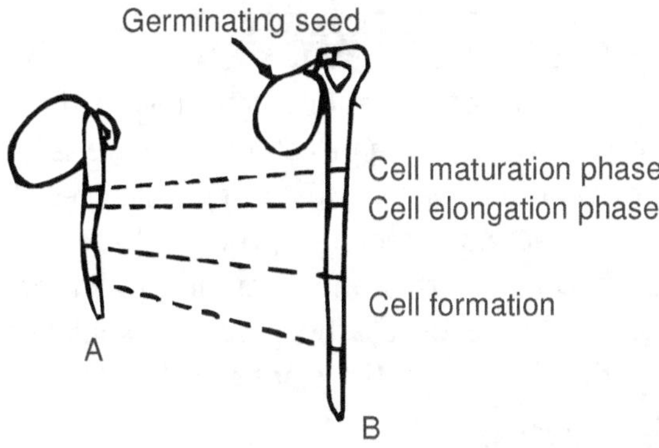

Fig. 5.1: Phases of growth

When growth begins in an organism, it is continued until an organism or organ acquires it's permanent form. The total period required by an organism to grow and to acquire a permanent form is known as the *"grand period of growth"*. This term was used for the first time by the German Physiologist **Sachs** (1873).

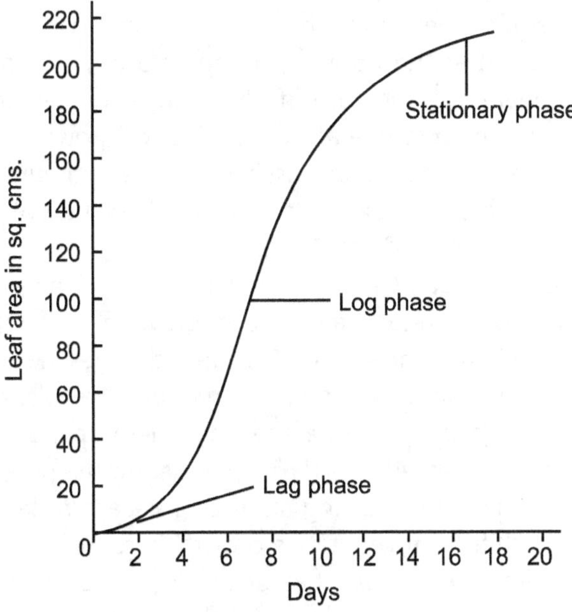

Fig. 5.2: Growth curve

5.3 FACTORS AFFECTING GROWTH

Growth is an effect of metabolic activities of the plant. When anabolic processes are more than catabolic, there is synthesis and accumulation of various organic substances which results in growth. There are various factors affecting growth. These factors can be broadly categorised as external and internal growth factors. Growth factors are considered to be all those sets of substances supplied externally or synthesised within the plant or cell.

5.3.1 External Factors

These factors mainly include climatic, edaphic and biological factors.

(a) Climatic factors: Climate is the atmospheric conditions of a particular region. It includes temperature, light, humidity, day length, wind, water and gases.

(i) Temperature: It is an important factor which affects the various metabolic activities. Growth of higher plants takes place in the range of temperature from 0°C to 45°C. Every plant requires specific temperature below which growth is not possible. Maximum growth takes place at the optimum temperature (28°C to 33°C). At each developmental stage, a different range of temperature is required. The optimum temperature required for growth differs from species to species. If temperature rises above the optimum level, rate of metabolic activities decreases and there is cessation of growth at maximum temperature (45°C to 50°C).

It is observed that plants growing in warm climate require a higher optimal temperature and those growing in cool climates require lower optimal temperature. Seasonal changes in temperature generally match with seasonal variation in plant growth (development). This is known as seasonal thermoperiodicity. Plant growth is also influenced by daily change in temperature. Low temperature at the early seedling stage is a necessary factor, which determines the shift from vegetative growth to reproductive growth. E.g. winter varieties of wheat, oat and rye are sown in the early autumn, so that the seedlings get exposed to low winter temperature and flower in summer. This low temperature treatment (0 - 5°C) is known as *vernalization*. In case of tropical plants, instead of chilling treatment, high temperature treatment is necessary for early flowering.

Low temperature reduces the rate of transpiration and lowers enzyme activities. Very low temperature causes chilling injury. High temperature increases rate of transpiration and photosynthesis and results in accumulation of organic compounds and increases growth. At very high temperatures (above 45°C) enzymes get denatured and growth stops due to destruction in metabolic activities.

(ii) Light: It is another important factor which has pronounced effects on the plant growth. Light quality, light intensity and duration of light affect growth in various ways. The various types of responses to light are phototropism, photoperiodism and photomorpho-genesis.

The range of visible light between 380 nm to 770 nm has physiological importance. Plants grow well in full spectrum of visible light. Light is necessary for photosynthetic activities. Rate of photosynthesis is highest in red light, while it is low in blue and yellow light. Thus, the light quality affects anabolic processes. In blue, violet light, internodal length increases and leaves expand. Red coloured light is favourable for growth. In dark, plant grows more rapidly, but mechanical tissues and pigments do not develop.

Light intensity has profound effects on photosynthesis, transpiration and indirectly on water and mineral absorption. Weak light intensity reduces photosynthetic rate, internodes become short and leaves are expanded. High intensity light causes increase in atmospheric temperature and heating of leaf surface, and results in increasing rate of transpiration. Very high intensity of light causes heat injuries, wilting of the leaves and reduces rate of growth.

Duration of light depends on the length of the day. Light period affects vegetative and the reproductive growth of the plant. The flowering response of the plant to light period is known as *photoperiodism*. On the basis of photoperiod required to induce flowering, plants are classified as long day plants, short day plants and day neutral plants. Under long day conditions, vegetative growth is profuse. Duration of light has an important role in inducing or suppression of flowering.

(iii) Humidity: Moisture content of the atmosphere is known as humidity. High humidity in the atmosphere reduces temperature and light intensity. Humidity between the 70 - 90% promotes growth. Low humidity (dry air) causes desiccation of the plant and reduces growth rate.

(iv) Wind: Strong wind current causes fluttering and bending of leaves. It increases rate of transpiration and creates water stress in plant which results in decreased growth rate. Steady wind is useful for increasing humidity around the plant.

(v) Water: Water is a major component of the protoplasm. It is necessary for the hydration of organic compounds, as a solvent for organic and inorganic compounds, agent for conducting minerals dissolved in it, for cooling, to maintain the pH and to maintain turgidity of the cell. Plant roots under field condition grow into moist soil, while the stems and leaves grow in a relatively dry atmosphere. Hence, for the growth of the roots and to maintain roots in an active state, the moisture level of the soil should be maintained at the optimum by providing water. Water only in liquid state is useful to the plant. Good rainfall is necessary for the healthy growth of crops. Vegetation can grow when sufficient water is available in the soil throughout the life span of the plant. Precipitation of water in the form of snow is not useful to the plant.

(vi) Gases: Atmosphere consists of various gases in various proportions. Gases like carbon dioxide, nitrogen and oxygen are necessary for the physiological processes in the plant. CO_2 is necessary for photosynthesis, O_2 is necessary for respiration and nitrogen is essential for synthesis of proteins, enzymes, hormones etc. Atmosphere contains 0.03% CO_2. If percentage of CO_2 is increased to 0.1%, the rate of photosynthesis increases, so plant shows vigorous growth. This technique is used in green house. It is observed that production of flowers and vegetables can be increased by 4 to 5 times by this technique. If percentage of CO_2 is increased beyond 0.15 %, it causes burning of shoot tips, un-controlled growth and less flower and fruit setting. Gases like SO_2, nitrogen oxides, fluorine (F), chlorine (Cl) and ozone (O_3) are toxic to plants. They retard the growth of the plant.

(b) Edaphic factors: Soil factors are known as edaphic factors. Soil is important for plants as it provides minerals, water and support to plants. Structure, texture and nutrient availability of the soil are the important edaphic factors. Soil structure means depth and nature of the top soil and subsoil layer. It is necessary for healthy growth of the root system. If depth of this layer is less, plant growth gets stunted. Soil texture means type of soil. Clay, loam, silt, clay loam, silty loam, and sandy loam soils are good for the growth of plants. These soils have higher water holding capacity. Soil texture directly influences soil water relationship, aeration and penetrability through it. It is also related to the nutrient status of the soil. e.g. Sandy soils are generally nutrient deficient, so not useful for the growth of the plant.

Nutrients necessary for healthy growth of the plants are broadly classified into two groups viz. macronutrients and micronutrients. Macronutrients are nitrogen, phosphorus, potassium, calcium, magnesium and sulphur. Micronutrients are iron, manganese, boron, molybdenum, zinc, copper, sodium, cobalt and silicon. Deficiency of any mineral element causes various deleterious effects on plants. If they are present in levels higher than normal, they change the pH of the soil and become toxic to the plant. So, for healthy growth of the plant, all minerals should be present in the soil, in the required amount and in the soluble forms.

(c) Biological factors: It includes various weeds, insects and soil micro-organisms such as blue green algae (BGA), nitrogen fixing bacteria, denitrifying bacteria, decomposing bacteria, mycorrhiza and soil fauna.

Weeds compete with crop plants and exhaust soil nutrients, so they negatively affect the growth of the crop. Many insects cause damage to crop plants as they destroy the growing apices and plant parts and spread many diseases. Micro-organisms growing in the soil increase soil fertility by adding nitrogen, by providing soluble phosphorus and by adding humus. Humus increases water holding capacity of the soil and make the soil loose and porous. Symbiotic bacteria and mycorrhizae provide nitrates and phosphates to plants. So, plants show vigorous growth. Animals like earthworms and crabs make the soil porous, which increases aeration and water holding capacity of the soil due to which roots grow vigorously.

5.3.2 Internal Factors

Growth and development of the plant is controlled by various internal factors. These factors are directly or indirectly controlled by the genetic constitution of the plant. They are broadly classified into three categories:

(i) Genetic factors

(ii) Nutritional factors

(iii) Hormonal factors

(i) Genetic factors: Every plant has a specific genetic constitution which determines the physical setup of the plant, so the dwarf variety remains dwarf and tall variety becomes tall. It also determines the branching and growth pattern, the season and period of vegetative and reproductive growth. Genetic factors are responsible for growth pattern of that plant.

(ii) Nutritional factors: Plants are autotrophic, so by the process of photosynthesis plant synthesizes various organic compounds. When a particular level of nutrition is achieved by the plant, it shows growth. When the rate of respiration is more than the rate of photosynthesis, there is utilization of organic compounds, so growth is not possible. When sufficient amount of food materials are accumulated and rate of photosynthesis is more than rate of respiration, then growth occurs. Other factors such as pigment contents, enzyme activity and capacity to store food reserves also affect the growth of the plant.

Photosynthetic rate: *The rate of CO_2 assimilation in plant is known as photosynthetic rate.* It depends on light quantity, quality and light interception. Light interception depends on leaf size, leaf area and total leaf canopy. CO_2 assimilation also depends on stomatal type and density. Other factors like CO_2 concentration, temperature, leaf pigments, activity of photosynthetic enzymes also control the photosynthetic rate. The photosynthetic rate of C_3 plants is less than the photosynthetic rate of C_4 plants. The rate of photorespiration is also very important as it governs the accumulation of carbohydrates during photosynthesis. Thus, photosynthetic rate is responsible for accumulation of food, which then affects the growth of the plant.

Pigment contents: As mentioned above, the rate of photosynthesis mainly depends on photosynthetic pigments like chlorophyll-a, chlorophyll-b, carotenes and xanthophylls. Chlorophylls are the master molecules in photosynthesis, hence the contents of chlorophylls indirectly affect growth. When the leaves become senescent or yellow, there is tremendous loss or degradation of chlorophyll contents; such plants do not show any growth, even when other factors are favourable. Chlorophyll a/b ratio is a crucial factor for efficient photosynthesis and capacity of the plant to harvest light. These pigments form photosystem I and photosystem II i.e. PS I and PS II which are very important in photosynthesis.

Enzyme activity: All the metabolic activities are governed by enzymes and hence any physiological process will be affected by enzyme activity. Their stimulation or inhibition will stimulate or inhibit processes like growth.

(iii) Hormonal Factors: Like genetic factors and nutritional factors the hormonal balance in plants affects their growth. The normal synthesis and accumulation of phytohormones like auxins, gibberellins and cytokinins control the plant growth. Hyper auxinity and excessive synthesis of gibberellins cause abnormal growth leading to excessive increase in plant height, size of leaves, flowers and fruits.

Plant growth is governed by the hormones, their concentration, biosynthesis etc. The natural hormones like Auxins, Gibberllins, Cytokinins, Abscisic acid and Ethylene are crucial in plant growth. Phytohormones even applied externally they contribute to growth of plants.

5.4 PLANT GROWTH REGULATORS (PGRS)

5.4.1 Introduction

Growth in plant is influenced by various naturally occurring organic substances other than nutrients. These substances are synthesised within the plant body and required in very less quantity. These substances are produced in one organ or tissue and transported to other sites where they produce specific effects on growth and development. They control various physiological processes and regulate growth, hence they are named as *growth regulators* or *plant growth substances (PGRs).* Another very common term 'Hormones is used by **Beylis** and **Starling** (1902). To distinguish them from animal hormones, they are named as *'phytohormones'.* They are formed in certain organs or tissues of the plant and translocated to other sites where they stimulate various morphological, physiological and biochemical responses.

5.4.2 Definition

According to different workers, 'phytohormones' or growth regulators are defined variously,

(1) *Growth regulators (plant hormones) are the organic substances which are synthesised in particular cells and which are transferred to other cells, where in extremely small quantities influence specific physiological processes {Phillips (1971)}.*

(2) *Phytohormones are organic substances, which are naturally produced in plants, control growth or other physiological functions at a site remote from its place of production and active in extremely minute quantities {(Thimann (1948)}.*

Types of Growth Regulators:

Plant growth regulators are broadly classified into two categories (i) Natural growth regulators (ii) Synthetic growth regulators.

(i) Natural growth regulators: These are naturally produced in plants or they are naturally occurring in plants e.g. auxins, gibberellins, cytokinins, florigen, anthesin, vernalin, coumarin, dormins, ethylenes and abscisic acid.

(ii) Synthetic growth regulators: These are synthesised on a commercial level e.g. morphactins, malformins, maleic hydrazide, phosphon – d, alar 85, cycocell etc.

5.5 PRACTICAL APPLICATIONS OF AUXINS

(1) Root initiation: Root initiation takes place when plants are propagated by cuttings. Success of cuttings depend on the development of adventitious roots. When lower end of a cutting is treated with 0.1 – 1% (1000 ppm to 10,000 ppm) auxin solution, it initiates roots. IAA, IBA and NAA are used for inducing the rooting of cuttings in woody plants. Readymade rooting hormones available in the market are Seradix, Phyxidium, Keradix,

Planofix, Germinator etc. These hormones are also used for initiation of roots in air layering gooty.

(2) Induction of flowering: By application of a minute quantity of auxin, flowering can be accelerated or induced in many plants. In day neutral plants, auxins act as florigen. Auxins like NAA and IAA promote flowering in tobacco, cotton, litchi, pineapple etc. Auxins induce ethylene production, hence auxin application tends to be inhibitory to flower formation in most plants.

(3) Fruit setting : Application of minute doses of auxins like IAA, IBA, NAA and 4 – chlorophenoxy acetic acid (4 – CPA) can increase fruit setting in many plants like chilly, brinjal, papaya, cucurbits, tomato etc.

(4) Parthenocarpy: It is a phenomenon of development of fruits without fertilization. Auxins like NAA and IBA induce enlargement of ovary and thus plant produces seedless fruits or fruits with non-functional seeds e.g. cucumber, watermelon, brinjal, lady finger etc.

(5) Prevention of premature drop of fruits: It is observed that mature fruits of citrus, (lemon, orange, sweet lime), grape, apple, pear are found to drop before the time of commercial harvest. This is due to formation of abscission zone near the stalk of the fruit. By application of auxins like 2 – 4 D, IAA, IBA and 2, 4, 5 trichlorophenoxy propionic acid (2, 4, 5 – TPA) abscission zone formation can be prevented.

(6) Seed germination and dormancy: For increasing percentage of seed germination, seeds of many plants (papaya, custard, apple, lady finger, aster, etc.) are treated with solutions of IAA, IBA, NBA, and 2, 4-D. Some synthetic auxins like ammonium thiocyanate (2% solution) can break the dormancy in case of potato tubers and maleic hydraxide and 1-naphthalene acetic acid can inhibit the sprouting of potato tubers.

(7) Apical dominance: A complete or partial inhibition of initiation and development of lateral buds by actively growing apical region is known as *apical dominance*. This is due to formation of auxin like IAA in the apical region which suppresses the growth of lateral buds. Thus, by application of solution of IAA on lateral buds, lateral growth, branching can be prevented in the plant.

(8) Weed control: Some synthetic auxins control the growth and kill the plant. Hence, they are used to control weeds. They have a very selective action e.g. 2, 4 – D kills broad leaved herbaceous dicotyledonous weeds. Other chemicals like 'Dalapon or 2–2–Dichloropropionic acid kills grasses and other monocot weeds.

(9) Prevention of lodging: In case of crops like paddy, wheat, rye, there is elongation and softening of cells in the basal internodes, so there is lodging (falling down). It causes great loss to the farmers. It can be prevented by applying chemicals like naphthyl acetamide to the basal portion of the plant.

(10) Tissue and organ culture: Auxins are useful in tissue and organ culture (micropropagation). They are used in different culture media for inducing rooting and differentiation in callus. They also promote cell division, so used for callus culture. Mostly auxins like IAA along with a cytokinin such as kinetin are used.

(11) Changing sex expression: Application of auxins at the bud stage can change sex expression of flower. Male flower can be converted into female flower, which initiates fruit development. e.g. In papaya, female flowers are important for fruit yield.

Various auxins are also used for prevention of leaf fall. Indole 3-acetic acid (IAA) is the main limiting and controlling factor for phloem and xylem differentiation. It is observed that low auxin levels induce phloem formation whereas both xylem and phloem differentiation takes place at higher auxin levels.

5.6 PRACTICAL APPLICATIONS OF CYTOKININS

(1) Cell division: Cytokinins are necessary along with auxins for cell division and *in-vitro* growth. Cytokinins inhibit root growth as well as stem elongation. Hence, they are used in tissue culture for culturing callus.

(2) Tissue differentiation: At low concentrations of cytokinin, tissue remains as undifferentiated callus. In tissue culture, bud formation and shoot initiation depends on higher concentration of cytokinins. When cytokinins and auxins are used in various ratios they initiate buds and shoots on the callus. Cytokinins also regulate chloroplast (plastids) formation.

(3) Retardation of senescence: Due to ageing, plants become yellow in colour and rapid break down of proteins takes place. It is known as *senescence*. It depends on physiological age of plants. When leaves are treated with kinetin they remain green for a longer period. It has practical significance in maintaining fruits and vegetables fresh and green for longer period so that they reach the market in a good condition.

(4) Breaking dormancy: Cytokinins are useful in breaking the dormancy of seeds and other plant organs. Dormancy of seeds of plants like tobacco, *Lactuca sativa* can be broken by treating them with Kinetin. Seeds of *Striga* can be induced to germinate in absence of frost by treating them with Kinetin.

(5) Apical dominance: Cytokinins when applied on lateral buds are able to release them from apical dominance.

(6) Mobility: Cytokinin affects mobility of organic substances. It is observed that amino acids, phosphates and various other organic substances get accumulated at the site of application of cytokinins.

(7) Protein synthesis: When a plant part is treated with kinetin there is increase in the protein content of the part due to increase in the rate of protein synthesis.

(8) Stomatal movement: When the whole leaf is treated with cytokinin there is increase in size of stomatal aperture, hence rate of transpiration also increases.

Cytokinins are also responsible for producing bisexual flowers on a male flowering plant. It is also reported that fruit set and fruit size in grape varieties and induction of parthenocarpy in fig is due to application of cytokinins.

(9) Morphogenesis: Cytokinin is very important in the formation of organs (Skoog and Miller, 1957).

(10) Callus formation in tissue culture is induced by cytokinin.

(11) Cytokinin can alleviate symptoms of stress conditions.

5.7 PRACTICAL APPLICATIONS OF GIBBERELLINS

(1) Seed germination: Gibberellins are useful for enhancing germination in case of seeds requiring cold treatment or light. In case of cereal grains, starch and other food reserves present in the endosperm undergo conversion into simpler compounds due to enhanced activity of enzyme amylase by the action of gibberellin. Thus, germination percentage of the seed increases.

(2) **Breaking dormancy and stimulating growth of dormant buds:** When gibberellin solution of particular concentration is applied to dormant buds of evergreen deciduous trees, buds will sprout. Thus, GA application is able to overcome bud dormancy by acting as a substitute for low temperature, long day or red light.

(3) **Rooting:** Gibberellins inhibit root initiation and the stimulation of rooting caused by auxins is also counteracted. So, cuttings are not treated with gibberellins.

(4) **Stem elongation:** When gibberellins are applied to plants they induce elongation of stem tissues and results in producing tall plants. This is due to cell elongation caused by gibberellins. Gibberllins also activate apical meristem.

(5) **Fruit setting and growth:** Gibberellic acid is commonly used by farmers on seedless grape varieties to increase the size and quality of the fruit. When GA is used along with auxin, it induces fruit set in case of deciduous trees such as apple, pear, citrus etc.

(6) **Leaf expansion:** In case of leafy vegetables such as cabbage, spinach, *Trigonella* and betel vine, GA solutions when sprayed on leaves causes increase in leaf area and hence productivity increases.

(7) **Flowering:** Gibberellic acid treatment promotes flowering in long day plants under short day condition. Hence, fruits and flowers can be made available in months that are not the usual time for their availability.

(8) **Pollen germination:** Gibberellic acid treatment increases percentage of pollen-grain germination and also accelerates the growth of pollen tube. This results in increased chances of fertilization.

(9) **Parthenocarpy:** When a solution of gibberellic acid is applied with a brush to the ovary it stimulates the growth of the ovary and induces parthenocarpy in case of brinjal, guava, and tomato etc.

(10) In floriculture, gibberellins are used to induce flowering and also to increase size of flowers in plants such as *Gerbera*, Roses, *Chrysanthemum* etc.

5.8 PRACTICAL APPLICATIONS OF ETHYLENE

(1) **Ripening of fruits:** Ethylene when applied as a foliar spray accelerates fruit maturity and induces uniform ripening in pineapple, fig, grape and many other horticultural plants including Mango, citrus, ber, sapota, litchi, strawberry, apple, pomegranate, etc. It also enhances fruit colour and yield.

(2) **Change in sex expression of flowers:** Ethylene when applied at three leaf stage increases the number of female flowers in cucumber, squash and melon.

(3) **Induction of flowering:** Application of ethylene is very useful for inducing flowering in many horticultural plants and ornamental plants.

(4) **Apical dominance:** Application of ethylene is found to be effective in some plants for inducing apical dominance and thereby it checks vegetative growth.

(5) **Induction of abscission and senescence:** Spraying of ethylene stimulates processes like abscission and senescence. This property is used in harvesting.

For example, cotton plants are sprayed with ethylene, the leaves fall down and then the mature cotton balls can be easily harvested. Even in many other plants, ethylene if sprayed two weeks before harvest, accelerates abscission of fruits and facilitates mechanical harvesting. The treatment of ethylene is also very useful to induce leaf fall in horticultural plants during bahar treatment, if there is some problem in natural leaf fall even after withholding water. This may be used for leaf, flower and fruit thinning.

(6) **Increasing the shelf life of fruits:** Fruits like apple, banana, papaya, strawberry, guava, sapota, ber, custard apple, and many others can be stored for longer time in an unripe condition by blocking the formation of ethylene in them. The shelf life of such perishable fruits can be increased by storing these fruits in ethylene depleted, CO_2 rich atmosphere.

(7) The opening of flowers of 'morning glory' in the early morning and their closing in the afternoon depends on ethylene concentration. The opening of carnation buds is accelerated by ethylene treatment while it inhibits the opening of rose buds at the same concentration.

(8) Ethylene is widely used to delay the ageing process in plants. Treatment with CO_2 or absorption of endogenously produced ethylene with mercuric perchlorate is used generally for this purpose.

(9) Ethylene is used to induce tolerance in plants against drought stress, flooding stress or temperature stress. It's application induces disease resistance in plants and protects the plants from the attack of insects and pathogens.

(10) Application of ethylene is helpful to increase the growth and thickness in plants. This increases sturdiness of the plant, and makes it resistant to mechanical injury.

(11) Ethylene causes growth and differentiation of shoot and roots in treated plants. Root growth is stimulated at low concentrations and inhibited at higher concentrations. Ethylene treatment induces enormous production of root hairs.

(12) Ethylene stimulates the formation of adventitious roots, leaves and stem. Ethylene induces rooting of cutting in nursery development.

5.9 PRACTICAL APPLICATIONS OF ABSCISIC ACID

(1) **Bud dormancy:** ABA induces bud dormancy in many plants and acts as growth inhibitor. The concentration of ABA is very high in dormant seeds. It is proposed that ABA is involved in the induction and maintenance of dormancy. ABA causes formation of resting buds in many woody plants. Thus, ABA is used to control the sprouting of buds in some plants like onions and tubers.

(2) **Senescence:** ABA promotes leaf senescence in a variety of plants. It is used to induce leaf fall in horticultural plants during 'bahar treatment', if there is no natural leaf fall after water stress. The purpose of bahar treatment is to induce dormancy or resting period in fruit plants. Hence, after the leaf fall the plants enter into a resting period.

(3) **Abscission:** ABA has the tendency to cause abscission of leaves, flowers and fruits. This can be practically used for thinning of leaves, flowers and fruits. Sometimes excessive leaves, flowers or fruits are formed, which are not necessary or are actually harmful to the plants. Therefore, to maintain the required number of leaves, flowers or fruits per plant, ABA spraying will be helpful.

(4) **Growth inhibition:** ABA is antagonistic in action to auxins, gibberellins and cytokinins and hence it inhibits growth in plants. ABA can be practically used for dwarfing the plants. It is also useful to check the excessive vegetative growth of plants, when it is not desired.

(5) **Fruiting and flowering:** ABA induces flowering in some plants like strawberries during long day conditions. ABA also induces flowering in short day plants. In grapes, ABA hastens ripening and colouring of fruits. ABA application in very low concentrations has a slight promoting effect on flowering. High concentration of ABA application inhibits or delays the flowering in many plants.

(6) **Stress resistance:** ABA has great importance in protecting the various types of plants during water or drought stress, salt stress, high temperature stress and cold injury. ABA induces stress tolerance in plants by altering nuclear gene expression or by forming "heat shock proteins" in plants. More ABA accumulation has been reported in all stress resistant plant species. It has been observed that external application of ABA causes stomatal closure and checks transpiration. Thus, it is involved in water conservation mechanism.

(7) **Root geotropism:** The geotropism in roots is due to ABA.

POINTS TO REMEMBER

- Growth can be defined as "Permanent and irreversible increase in an organism". Or "Growth is a vital process which brings about a permanent change in any organism or it's part with respect to it's size, weight and volume."

- Growth in plants is due to a special tissue known as 'meristem' which is not present in animals.

- The growth in plants is restricted to certain regions only, which are known as growth regions.

- The qualitative changes take place in plant body such as germination of seeds, formation of flowers and fruits, falling of leaves etc. is known as development in plants.

- The study of development from a morphological point of view is known as morphogenesis.

- The total period required by an organism to grow and to acquire a permanent form is known as the "grand period of growth".

- The flowering response of the plant to light period is known as photoperiodism.

- Soil factors are known as edaphic factors

- The rate of CO_2 assimilation in plant is known as photosynthetic rate.

- Growth regulators (plant hormones) are the organic substances which are synthesised in particular cells and which are transferred to other cells, where in extremely small quantities influence specific physiological processes {(Phillips (1971)}.

- Plant growth regulators are broadly classified into two categories (i) Natural growth regulators and (ii) Synthetic growth regulators.

- Natural growth regulators are naturally produced in plants. e.g. auxins, gibberellins, cytokinins, florigen, anthesin, vernalin, coumarin, dormins, ethylenes and abscisic acid.

- The plant-growth regulator (PaRs) are playing very important role in plants and they are involved in the processes like seed germination, root and stem initiation, flavouring, fruiting etc.

- They are involved in parthenocarpy avoiding premature dropping of flowers and fruits.

- They are very important in senescence, fruit ripening, increasing shelf life of fruits, Protein synthesis, etc.

EXERCISE

1. Define growth and describe the kinetics (phases) of growth.
2. Explain the growth curve with suitable diagram.
3. Explain the impact of humidity, wind and water on growth.
4. Discuss the role of edaphic factors on growth.
5. How do biological factors affect plant growth?
6. Explain the impact of genetic and nutritional factors on growth.
7. What are plant growth regulators? Give their general applications.
8. Give an account of practical applications of gibberellins.
9. Give practical applications of auxins.
10. Give an account of the practical applications of cytokinins.
11. Give an account of the practical applications of abscisic acid.
12. Give an account of practical uses of ethylene.
13. Describe the role of auxins in plants.
14. Comment on the role played by GA and IAA in plant growth.
15. Explain the role of ethylene in ripening of fruits.
16. Distinguish between auxins and gibberellins, auxins and cytokinnins, gibberellins and abscisic acid, auxins and ethylene.
17. Explain the role of plant growth regulating substances (PGRs) in agriculture, floriculture and horticulture.
18. Enlist any two plant growth regulators involved in development of partherocarpic fruits.
19. Enlist the growth hormones involved in change in sex expression in plants.
20. Which hormones induce flowering?
21. Name the Plant growth regulators:
    (a) Inducing root inition, fruit setting.
    (b) Inducing parthenocarpy.
    (c) Retaration of senescence and stomatal movement.
    (d) Inducing stem elongation, flowering and fruiting.
    (e) Inducing fruit ripening increasing shelf life of fruits.
    (f) Inducing bud dormancy senescence and stress tolerance.

STRUCTURE OF PLANT CELL

◆ POINTS TO LEARN ◆

6.1 Introduction
6.2 Structure of Plant Cell
6.3 Prokaryotic Cell
    6.3.1 Difference between Prokaryotic Cells and Eukaryotic Cells
*** Points to Remember**
*** Exercise**

6.1 INTRODUCTION

Living world consists of plant kingdom and animal kingdom. More than 8.7 million species are living on this planet Earth. All the living organisms e.g. plants, animals and even microorganisms are made up of cells. They may be single celled e.g. bacteria, a green algae like **Chlorella**. While higher plants as well as animals are multi-cellular. The cells are the basic structural and functional units of all plants and animals. The cells provide definite size, shape and structure to all the living organisms. Not only this but the cells in plant and animal body carry out different types of functions and keep them active and alive. The cells are of two types: (i) Eukaryotic, (ii) Prokaryotic.

The organisms having eukaryotic cells are known as Eukaryots and the others as Prokaryots. Mostly the plant and animal cells are eukaryotic.

Each cell of eukaryotic organisms contain different functional structures which are collectively known as cell organelles. These are found only in eukaryotic cells. While the bacteria and archaea like prokaryotes do not have any type of cell organelles. The cell organelles are membrane bound, having distinct structures and functions.

6.2 STRUCTURE OF PLANT CELL

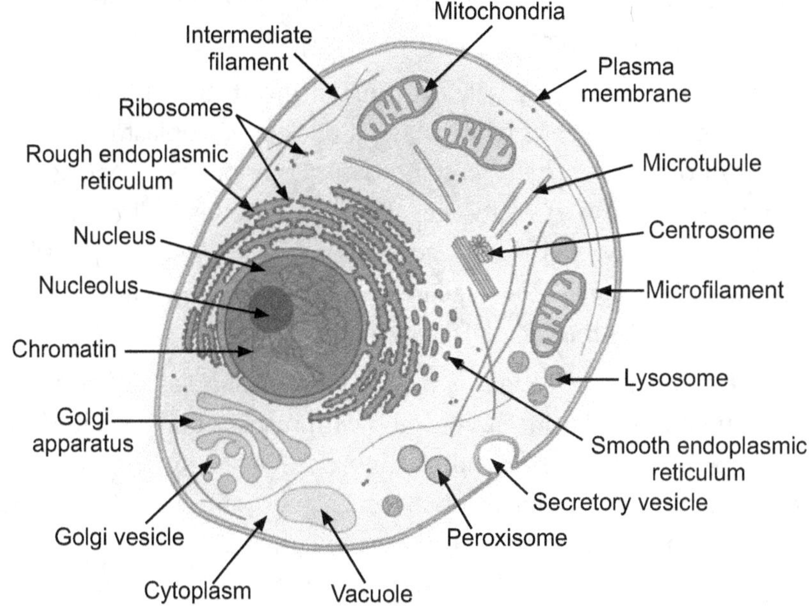

Fig. 6.1: Structure of Plant Cell (Eukaryotic)

Plant cell differs from animal cells by having distinct cell wall made up of cellulose, having chloroplast and central vacuole. Plant cell lacks centriole and cilia/flagella. Let us know about the cell structure and cell organelles of plant cell (Fig. 6.1).

(i) Cell wall: It is relatively thick and made up of cellulose, hemicellulose and pectin. The cell wall is protective layer of every cell providing structural support and protection to plant cell. The rigid, thick layer of cell wall is found outside the plasma membrane surrounding the cell. Besides the cellulose it also contain proteins and some polysaccharides. The pores found in the cell wall allow water and nutrients to move in and out of the cell. The cell wall also protects the cell from bursting when water enters in it. This cell wall is always known as primary cell wall but some cells may have secondary cell wall made up of lignin.

(ii) Plasma membrane: Every plant cell is enclosed by plasma membrane. It is also known as cell membrane or cytoplasmic membrane. It is made up of lipid bilayers and proteins. Plasma membrane is selectively permeable allowing the entry and exit of selective molecules according to the cell requirements.

(iii) Cytoplasm: It is a jelly like sustance distributed throughout the cell between the cell membrane and nucleus. Cytoplasm is mainly composed of water, organic and inorganic compounds. Cytoplasm is the most essential part of the cell in which different cell organelles are embedded. Cytoplasm also contains many enzymes responsible to carry various metabolic reactions. Its main site for most of the chemical reactions that take place in cell.

(iv) Nucleus: The nucleus is made up of nucleoplasm, nucleolus, nuclear membrane or nuclear envelop with nuclear pores on it. Plant cell has eukaryotic/true nucelus having double membrane. It is the largest cell organelle which functions as the control centre of the cellular activities and containing the genetic material of cell i.e. DNA. Nucleus is a dark, rounded body surrounded by nuclear membrane.

It is porous and acting as a wall between cytoplasm and nucleus itself. Within the nucleus there are tiny sperical bodies called as nucleolus, which carries chromosomes. The hereditary characters carried by genes and they are present on chromosomes. Nucleus controls the characters and functions of a cell.

(v) Central vacuole: Most of the plant cells have central vacuole, which is surrounded by a membrane known as tonoplast. It occupies generally 30% of cell volume and sometime upto 90% of cell volume. The vacuoles are very common in arenchyma cells of aquatic plants. Vacuoles are mostly defined as storage bubbles of irregular shapes distributed in the cell. They are fluid-filled organelles enclosed by membrane.

The vacuoles contain cell sap which is a mixture of enzymes, ions, salts etc. It sometimes store toxic substances and protects the cells. The vacuoles also store food and variety of nutrients required for the survival of cell.

(vi) Endoplasmic reticulum: It is network of membranous canals filled with fluid. The ER is a transport system of the cell, involved in transporting materials throughout the cell.

There are two different types of ER

(a) Rough Endoplasmic Reticulum (RER).

(b) Smooth Endoplasmic Reticulum (SER).

(a) Rough Endoplasmic Reticulum (RER): It is composed of cisternae, tubules and vesicles found throughout the cell. As it is having many ribosomes on it like a rough structure and hence known as rough ER. Its main function is protein synthesis.

(b) Smooth Endoplasmic Reticulum (SER): These are without ribosome and hence called as SER. It is mainly involved in synthesis of lipid and steriods. It also plays a significant role in detoxification of a cell (Fig. 6.2).

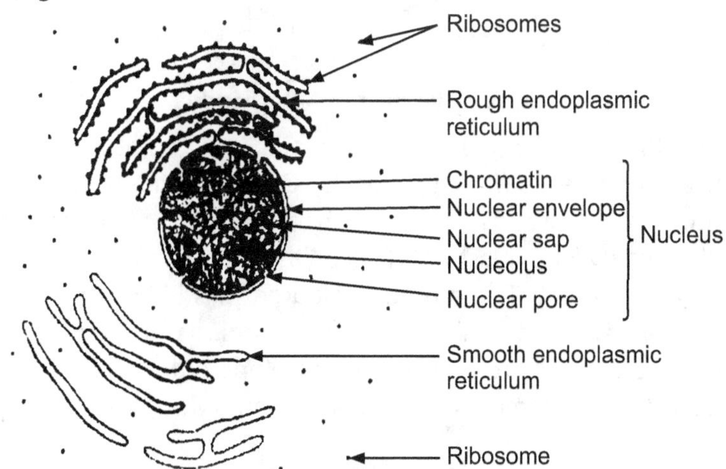

Fig. 6.2: Structure of endoplasmic reticulum

(vii) Mitochondria: These are of various shapes and size. The mitochondria are double membraneous structure, having outer and inner membrane. The outer membrane is smooth but inner membrane is folded in to cristae. The cristae are suspended in to matrix, which contain many enzymes of Kreb's cycle (aerobic respiration), DNA, ribosomes and especially ATP synthase particles ($F_0 - F_1$ particles). Mitochondria are known as "Power house of a cell" as it is mainly involved in the synthesis of energy molecules "ATP" from ADP + iP. The process is known as "Oxydative phosphorylation" ATP is the "cellular currency" (Fig. 6.3).

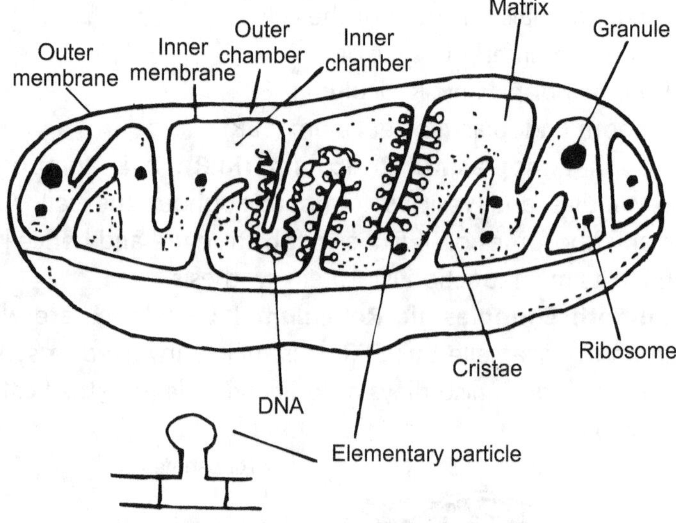

Fig. 6.3: Structure of Mitochondria

(viii) Plastids: Every plant cell contains large number of plastids, which are of three different types (a) Green plastids – Chloroplast, (b) Coloured plastids (Other than green) – Chromoplast, (c) Colourless (White) plastids – Leucoplasts.

(a) Chloroplast: These green coloured plastids contain a green pigment known as chlorophyll. Chloroplasts is also double membraneous structure having two envelops outer and inner. Each chloroplast has stroma or matrix and grana which are made up of

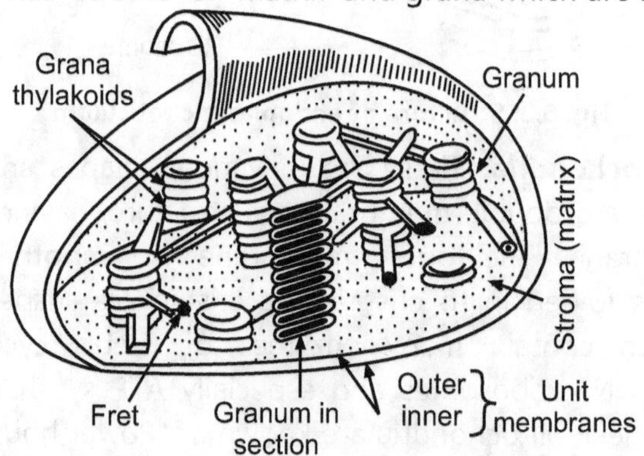

Fig. 6.4: Structure of Chloroplast

thylakoid membranes put together like a pile of coins. The light reaction of photosynthesis takes place in grana – which release the by-product like O_2 and energy molecules like ATP. Stroma or matrix is actively engaged in dark reaction of photosynthesis in which CO_2 is fixed and reduced to glucose with the help of water and energy (ATP). Thus, chloroplasts are solar energy harvesting molecules, which convert solar energy into useful chemical/biological energy with the release of oxygen in the atmosphere (Fig. 6.4).

Chloroplast also contain double stranded circular DNA, '70S' ribosomes and enzymes required for carbohydrate synthesis.

(b) Chromoplasts: It includes fat soluble carotenoid pigments like xanthophylls, carotenes, anthocyanins etc. which impart beautiful colours to flower petals, fruit walls, varigated leaves, seeds etc. These helps in pollination and fruit as well as seed dispersal.

(c) Leucoplasts: These are colourless or white plastides found in underground parts like roots and tubers. They usually function as storage plastides. e.g. Amyloplasts storing starch in potato.

(ix) Ribosomes: These are non-membrane bound structures found in close association of endoplasmic reticulum. These are composed of 2/3rd of RNA and 1/3rd of proteins. They are named as 70s (found in prokaroytes) or 80s (found in eukaryotes). Both types of ribosomes are composed of two subunits. Ribosomes are either encompased within endoplasmic reticulum or freely traced in the cell cytoplasm. Ribosomal RNA and ribosomal proteins constitute ribosomes. The main function of ribosomes is protein synthesis (Fig. 6.5).

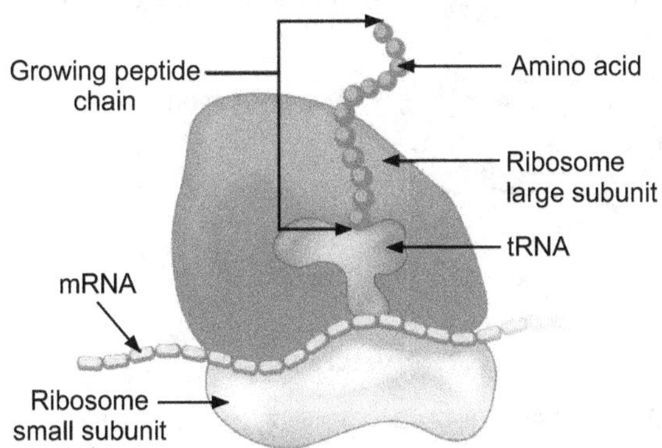

Fig. 6.5: Structure of Ribosome

(x) Golgi apparatus: It is also known as Golgi bodies or Golgi complex. It is membrane bound structure composed of series of flattened stacked pouches called cisternae. The main function of Golgi bodies is transportation, modification and packaging of proteins and lipids to targated destinations. Golgi apparatus is found within the cytoplasm (Fig. 6.6).

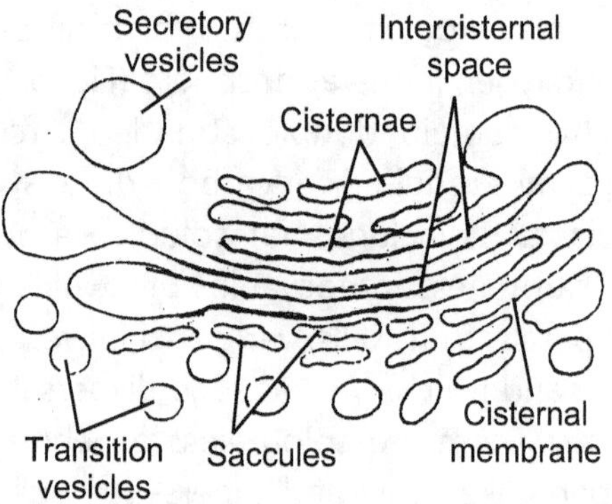

Fig. 6.6: Structure of Golgi apparatus

(xi) Peroxysomes: These are spherical organelles highly important in lipid destruction. They mainly contain oxydative enzymes and hence protect the cells from Reactive Oxygen Species (ROS). It is also a site for breakdown of fatty acids.

(xii) Lysosomes: These are involved in the destruction of proteins. These are bag like structures containing many hydrolysing enzymes. It is a cell's recycling centre (Fig. 6.7).

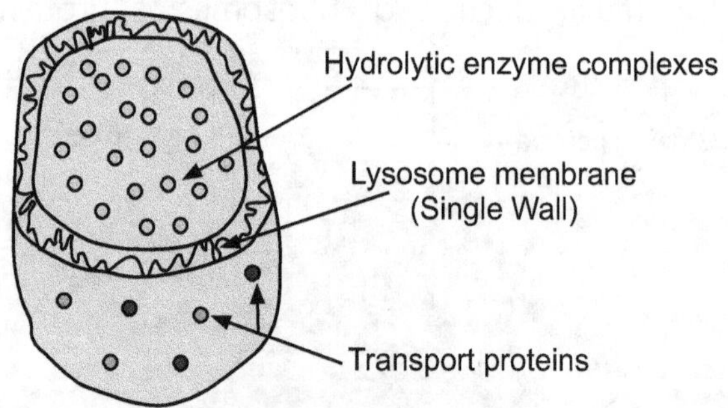

Fig. 6.7: Structure of lysosome

(xii) Nucleotides: There are two types of nucleotides i.e. DNA (Deoxyribo Nucleic Acid) and RNA (Ribo Nucleic Acid) present in the cells. DNA is usually double stranded and RNA is single stranded. The previous one carrying genetic material and the next one playing role in protein synthesis. DNA found in different cell organelles and nucleus.

Cytoskeleton: It is a continuous network of filamentous proteinaceous structures that run throughout the cytoplasm, from the nucleus to the plasma membrance. The cytoskeleton matrix is composed of different types of proteins that can divide rapidly or disassemble depending on the requirement of the cells. Its primary function is to provide definite shape and mechanical resistance to the cell against deformation. It also helps in motility and during cytokinesis.

6.3 PRAKARYOTIC CELL

Prakaryotic Cell: This term is derived from Greek word Pro-primitive, karyon = nucleus. The prokarytic cells are most primitive, found in bacteira, blue-green algae and archaea. As the name indicate the prokaryotic cell lacks a defined nucleus. Instead of nucleus they have nucleoid. It is a irregularly shaped region where the genetic material is localised. This cell lacks organells or other internal membrane bound structures.

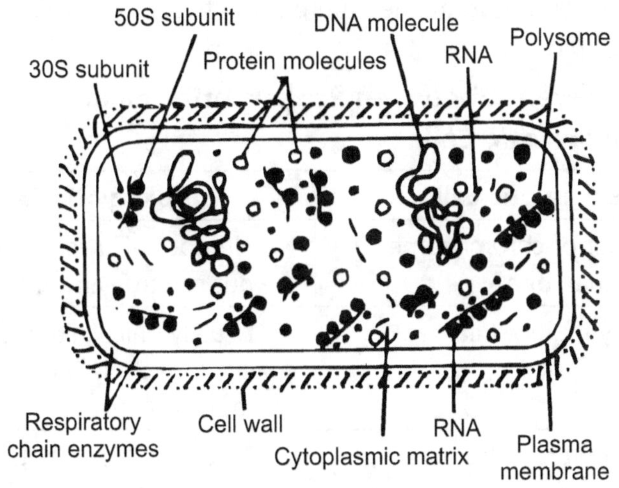

Fig. 6.8: Prokaryotic Cell

The prokaryotic cell has following important structures.

(i) Cell wall: The prokaryotic cell shows outer cell wall composed of carbohydrates, lipids, proteins, muramic acids, organic salts etc.

(ii) Capsule: The outer slimy layer of polysaccharides called capsule.

(iii) Plasma membrane: Cytoplams of the prokaryotic cell is enveloped by thin plasma membrane which is composed of lipid and protein. It also contain respiratory chian enzymes, multi enzymes complex. It also shows the presence of desmosomes and mesosomes.

(iv) Cytoplams: It is dense, colloidal in nature containing glycogen, proteins and fat granules. It is without cell organells like E.R. mitochandria plastides, lysosomes, but in some bacteria bacteriochlorophyll is present. In the center there is nucleoid sharing presence of single, large circular double stranded DNA molecule.

(v) Flagella: many prokaryotic cells have flagella or cilia with which they swim. The flagella are whip like and many in number. Some bacteria show hair like out growth on their bodies which are known as pilli or fimbriae.

The structure of eukaryotic cell is described in detail in previous pages. Now we will study the difference between the prokaryotic and eukaryotic cell:

6.3.1 Difference between Prokaryotic Cell and Eukaryotic Cell

Eukaryotic Cell	Prokaryotic Cell
1. Number of chromosomes more than one.	1. Number of chromosome one but not true chromosome.
2. Usually multicellular.	2. Usually unicellular (some cyanobacteria may be multicellular).
3. Tue membrane bound nucleus present.	3. True membrane bound nucleus absent.

Contd...

4. Genetic transformation by meiosis and fusion of gametes.	4. Partial, un-directional transfer of DNA.
5. Lysosomes and peroxisomes present.	5. Lysosomes and peroxisomes absent.
6. Endoplasmic reticulum present.	6. Endoplasmic reticulum absent.
7. Mitochondia present	7. Mitochondia absent
8. Cytoskeleton present	8. May be absent.
9. Microtubules present.	9. Microtubules absent or rare.
10. Eukaryotes wrap their DNA around proteins called as histones.	10. Histones absent.
11. Ribosomes larger 80s type.	11. Ribosomes smaller 70s type.
12. Chloroplast present (plants)	12. Chloroplast absent.
13. Permeability of nuclear membrane is selective.	13. Nuclear membrane not present.
14. Plasma membrane with steroid.	14. Not with steroid.
15. Cell wall present.	15. Cell wall absent.
16. Cell size 10 – 100 μm.	16. Cell size 1 – 10 μm.
17. Mitosis occur	17. Mitosis does not occur.
18. Histones present.	18. Histones absent.
19. Respiratory system in Mitochondria.	19. Respiratory system in parts of plasma membrane because mitochondria absent.
20. Cytoplasmic movement by streaming.	20. Cytoplasmic streaming absent.
21. Example: Plant cell and animal cell.	21. Example: Bacterial cell.

POINTS TO REMEMBER

- Cells are mainly of two types (a) Eukaryotic (b) Prokaryotic.
- Eukaryotic cells are present in plants and animals.
- Prokaryotic cells are present in bacteria and BGA.
- Eukaryotic cell is evolved one while prokaryotic cell is highly primitive.
- Eukaryotic cell has cell wall, plasma membrane cytoplasm, nucleus, nucleolus and nuclear membrane, chromosomes etc.
- The eukaryotic cell has many cell organells like chloroplast, mitochondira, golgibodies, endoplasmic retitculum, lysosomes, peroxisome etc.
- The cell organelles may be membranous or without membranes.
- Chloroplast during photosynthesis prepare food from $CO_2 + H_2O$ in presence of sunlight releasing oxygen.
- Mitochondia is the power house of cell involved in ATP synthesis.
- ER are of two types smooth and rough. These are actively important in synthesis of proteins as well as lipids and transport of many materials.
- Prokaryotic cells do not have ER, Mitochondia, Golgibodies etc.
- Prokaryotic cells have nucleoid, primitive nucleus containing single circular DNA strand.
- Eukaryotic cells do not have flagella or cili but prokaryotic cell have flagella cilli or pilli.

EXERCISE

1. Explain what is Eukaryotic cell?
2. Explain what is Prokaryotic cell?
3. Draw labelled diagram of plant cell.
4. Give an account of cell organells present in plant cell.
5. Give the functions of
   - (i) Cell wall,
   - (ii) Plasma membrane
   - (iii) Cytoplasm
   - (iv) Mitochondira
   - (v) ER
   - (v) Golgi bodies

6. Explain the role of
   (i)   Chloroplast   (ii) Golgi bodies
   (iii) Lysosomes   (iv) Vacuole
   (v) Cytoskeleton
7. Explain the functions of plant cell.
8. Distinguish between eukaryotic and prokaryotic cell.
9. Write short notes on:
   (i)   Plant cell
   (ii) Eukaryotic cell
   (iii) Prokaryotic cell
   (iv) Nucleus in plant cell
   (v) Nucleoid of Prokaryotic cell
   (vii) Golgi complex
   (viii) Cell wall in Plant cell
   (ix) Vacuoles of plant cell
10. Match the following:

1) Nucleus	1) Bacteria
2) Vacuoles	2) Selectively permeable
3) Oxysomes	3) Power house of cell
4) Mitochodria	4) Stress tolerance
5) Plasma membrane	5) Tonoplast
6) Flagella / Pilli	6) Controlling centre of a cell

11. Sketch and label the diagram of prokaryotic cell.

PLANT CELL WALL

◆ POINTS TO LEARN ◆

7.1 Introduction
7.2 Components of Primary Cell Wall
7.3 Structure and Functions
***   Points to Remember**
***   Exercise**

7.1 INTRODUCTION

In plant cell, the plasma membrane is always protected by a thick rigid cell wall made up of cellulose, hemicellulose and lignin. It provides additional strength and rigidity to the plasma membrane of plant cell. It is absent in animal cell and prokaryotic cells. Microtubules guides the formation of plant cell wall. Cellulose is laid down by enzymes to form the primary cell wall. Some plants also have secondary cell wall which contain lignin. The cell wall acts as filterating mechanism of plant cell. Cell wall of epidermis and endodermis contain suberin. Cell wall also show the presence of some enzymes like hydorlases, esterases, peroxidases and transglycolases. It sometimes also contains some minerals elements.

7.2 COMPONENTS OF PRIMARY CELL WALL

Plant cell wall is composed of three layers:
(i)   Middle lamella,
(ii) Primary cell wall,
(iii) Secondary cell wall.

Primary cell wall is the first formed cell wall situated closest to the inside of the cell. The primary cell wall is generally thin, flexible and extensible layer formed while the cell is growing. Primary means it is the growing cell wall.

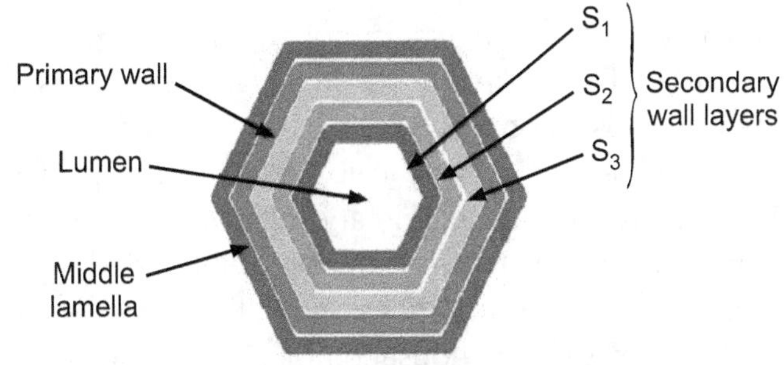

General plant cell wall layers

Fig. 7.1 (A): Structure of Primary Cell Wall

The middle lamella is pectin layer. It cements the cell walls of two adjoining cells together. It is the first layer formed during cell division. Primary cell wall of land plants is composed of the polysaccharides like cellulose, hemicellulose and pectin. Often other polymers such as lignin, suberin or cutin are anchored or embedded in it. Primary cell wall is formed after the middle lamella.

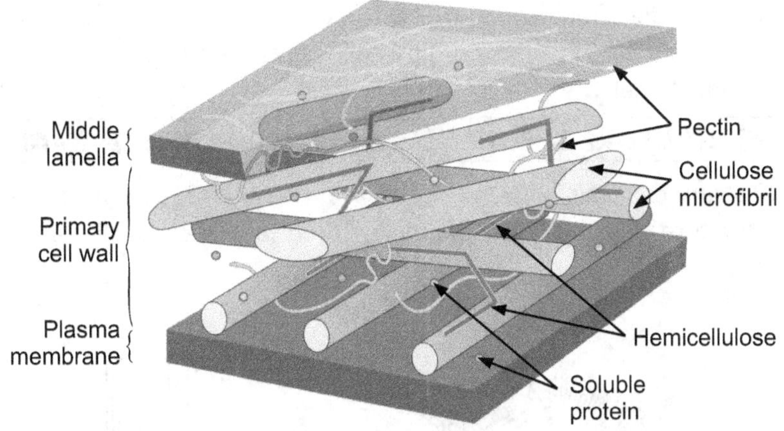

Fig. 7.1 (B): Components of Primary Cell Wall

Primary cell wall along with polysaccharides also contain lesser amount of structural glycoproteins, phenolic esters, and some enzymes. Primary cell wall along with celluose also shows the presence of xyloglucan, glucuronoxylan, arabinoxylan, glucomannan and galactomannan like hemicellusosic substances. Primary cell wall also shows the presence of different types of pectins like homogalacturonans, rhamnogalacturonans etc. The organization and interactions of wall components is not yet known perfectly. The cellulose microfibrills are linked via hemicellulosic fibres to form cellulose-hemicellulose network which is embedded in the pectin matrix. The outer part of primary cell wall is made up of wax and cutin, which is known as cuticle. Primary cell wall basically consists of cellulose fibres of great tensile strength embedded in a water saturated matrix of polysaccharides and structural glycoproteins. Cellulose consists of several thousands of glucose molecules linked end to end. Primary cell wall also shows presence of Matrix polysaccharides. These are of two types: Hemicelluloses and Pectic polysaccharides or pectins. Both are synthesized in golgi bodies and brought to the cell surface and secreted into the cell wall. Proteins: Very small amount of proteins are present in primary cell wall but they perform many functions. e.g. hydroxyproline rich glycoproteins, extensin proteins.

Formation of Primary Cell Wall:

The middle lamella is laid down first formed from the cell plate during cytokinesis and the primary cell wall is then deposited in side the middle lamella.

7.3 STRUCTURE & FUNCTIONS OF PRIMARY CELL WALL

1. Primary cell wall is always found in dividing and growing cells to structure of primary cell wall allow the cell wall expansion during growth.

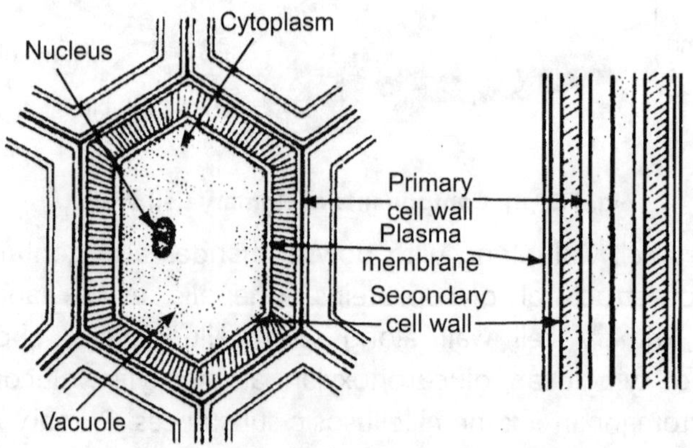

Fig. 7.2: Structure of Primary Cell Wall

2.　Primary cell wall is thinner and less rigid and helps for proper expansion of growing and dividing cells.

3.　A fully grown plant cell usually retains it's primary cell wall.

4.　Thinner primary cell walls are capable of serving a structural and supportive role only when the vacuoles within the cell are filled with water to the point that they exerts a turgor pressure against the cell wall.

5.　The turgor induced stiffening of primary cell wall protects the cells from bursting.

6.　If the turgor pressure of a cell is lost the flowers and leaves of plants show wilting.

7.　The primary cell wall of most plant cells is freely permeable to small molecules such as proteins. The pH is an important factor for the transport of molecules.

8.　Primary cell walls extend (grow) by a mechanism called acid growth mediated by expansions type of proteins.

9.　The outer part of primary cell wall of plant epidermis is usually impregnated with cutin and wax forming a permeability barrier known as cuticle. It is very protective in nature and minimize the excessive transpiration rate thereby avoiding the wilting of plants. It also avoids the fungal and bacterial infections. Because of cuticle the plants become water stress and heat stress tolerant.

10.　Plasma membrane of a cell is protected by primary cell wall.

11.　It provides additional strength and rigidity to plasma membrane.

POINTS TO REMEMBER

- Primary cell wall is first formed cell wall.
- It is deposited inside the middle lamella.
- Primary cell wall is thin, flexible and extensible layer.
- It protects the plasma membrane.
- Primary cell wall is made up of cellulose, hemicellulose and pectin.
- It also contains lesser amount of glycoproteins.
- The outer part of primary cell wall is made up of wax and cutin forming cuticle.

- Primary cell wall is always found in dividing and growing cells.
- It allows the cell wall expansion during cellular growth.
- Primary cell wall functions as structural and supportive device for the cell.
- Primary cell wall is permeable to small molecules like proteins.
- Primary cell wall also contain matrix polysaccharides and some special types of protein known as extensin proteins, glycoproteins.

EXERCISE

1. What is primary cell wall?
2. Explain how primary cell wall is formed.
3. What are the different components of primary cell wall?
4. Briefly explain the composition of primary cell wall.
5. Sketch and label the structure of primary cell wall.
6. Explain the major functions of primary cell wall.
7. Explain the chemical composition of primary cell wall.

8
CHAPTER

PLASMA MEMBRANE

♦ POINTS TO LEARN ♦

8.1 Introduction

8.2 Chemical Composition of Plasma Membrane

8.3 Membrane Models

    8.3.1 Lipid – Lipid Bilayer Model

    8.3.2 Bilayer Model

    8.3.3 Fluid Mosaic Model

8.4 Functions of Plasma Membrane

*** Points to Remember**

*** Exercise**

8.1 INTRODUCTION

The plasma membrane also known as cell membrane or cytoplasmic membrane or plasmalemma. It is a biological membrane that separates the interior of a cell from it's outside environment. A plasma membrane by definition is a fluid, phospholipid bilayer that plays a key role in many cellular processes. In short plasma membrane is the boundary between the cell and it's outer environment regulating the entry and exit of different materials, molecules or substances at it is selectively permeable. It is found in all eukaryotic cells of plants and animals.

8.2 CHEMICAL COMPOSITION OF PLASMA MEMBRANE

1. Phospholipids:

Plasma membrane is made up of phospholipid bilayer, which is two layers of phospholipid back to back. Phospholipids are lipids with a phosphate group attached to them. Phospholipids have one head and two tails. The head is polar and hydrophilic or water loving. The tails are non-polar and hydrophobic – or water fearing. Phospholipids are lined up in two layers with the heads pointing outward and tails hidden in the middle. The phospholipids are present on both the sides (inside and out side) of the membrane, hence the term "bilayer". In the plasma membrane phospholipids, glycolipids and sterols are present. Phospholipid is a lipid made up of glycerol, two fatty acid tails and a phosphate linked head group. Biological membranes are having two layers of phospholipid hence known as phospholipid bilayer.

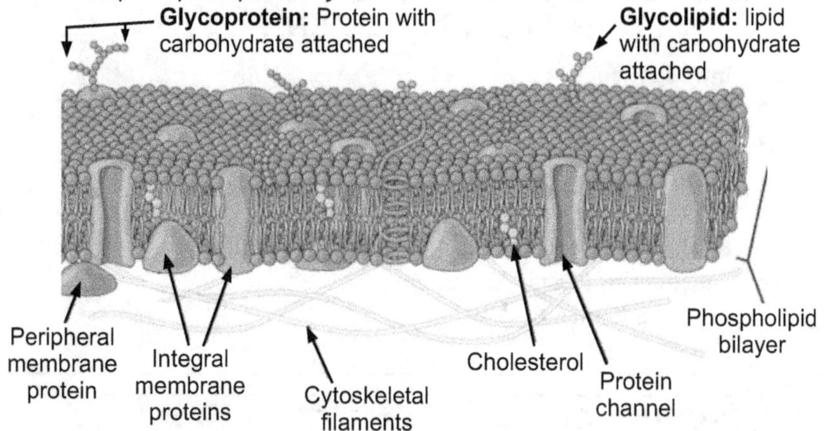

Fig. 8.1: Structure of Plasma Membrane

Phospholipids are the basic fabric of plasma membrane which constitute over 50% of lipids in plasma membrane. These are most suitable for the membrane because they are amphipathic i.e. they have both hydrophilic and hydrophobic regions. There are three types of lipids – **(i) Phospholipids, (ii) Glycolipids, (iii) Sterols.** Along with phospholipids, cholesterol and glycolipids are present. Glycolipids are oligosaccharides containing lipid molecule.

2. Plasma membrane Carbohydrates:

The carbohydrates are also present in the plasma membrane in the form of glycolipids and glycoproteins present on the external membrane surface. The oligosaccharide chains play important roles in cell to cell recognition process.

3. Plasma membrane proteins:

The active functions of plasma membrane are dependent on the proteins. The membrane proteins carryout the important functions such as (i) cell adhesion, (ii) energy transduction, (iii) signaling, (iv) cell reorganization, (v) transport.

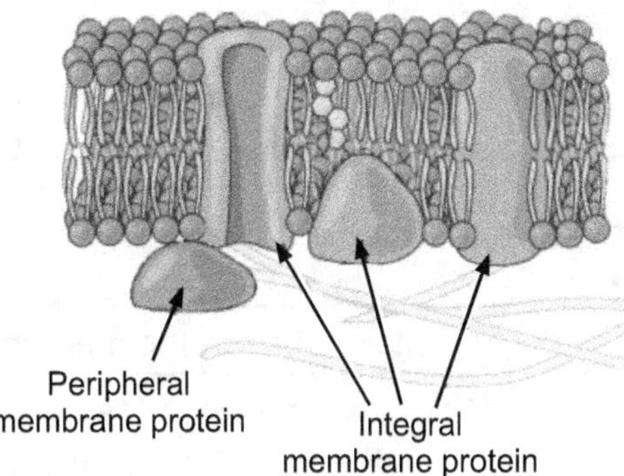

Fig. 8.2: Proteins in plasma membrane

The membrane proteins are of different types.

(i) Intrinsic or Integral membrane proteins: These are embedded in the hydrophobic region of the lipid bilayer. **Transmembrane proteins** are the classic examples of intrinsic proteins. These proteins form channels for the transport of substances in and out of the cell through plasma membrane.

(ii) Extrinsic or peripheral membrane proteins: These are associated loosely with the hydrophilic surfaces of the lipid bilayer by means of ionic bonds or calcium bridges. The spectrin and glyco proteins are the examples of intrinsic and extrinsic membrane proteins.

(iii) Spectrin: It is the most abundant type of protein which constitutes about 30% of the membrane proteins. These are long fibrous molecules.

(iv) Glycoproteins: The hexoses are bound to proteins forming glycoproteins. These are present on the outer surface of membrane. These proteins are mostly attached to oligosaccharides.

(v) Structural proteins: These proteins form the back bone of cell membrane.

(vi) Carrier proteins: They transport substances across the membrane against the concentration gradient.

(vii) Enzyme proteins: More than 30 different types of enzymes are found in plasma membrane. E.g. Mg ATP$_{ase}$, Na-K ATP$_{ase}$, alkaline phosphates. These enzymes are important in ion transfer across the plasma membrane.

Cell membrane proteins are mainly of three types: (1) Transport proteins, (2) Receptor proteins, (3) Recognition proteins.

The various components of the plasma membrane are summarized in the following table.

Table 8.1: The components of plasma membrane

Name of the Component	Location
1. Phospholipid	1. Main fabric of the membrane.
2. Cholesterol	2. Tucked between the hydrophobic tails of the membrane phospholipid.
3. Integral proteins	3. Embedded in the phospholipid bilayer. May or may not extend through both layers.
4. Peripheral proteins	4. Present on the inner or outer surface of phospholipid bilayer but not embedded in the hydrophobic core.
5. Carbohydrates	5. Attached to proteins or lipids on the extracellular side of the membrane (forming glycol proteins and glycolipids)
6. Cholesterol	6. Tucked between the hydrophobic tails of the membrane.

8.3 MEMBRANE MODELS

For the detailed study of physical and biological features of plasma membrane two types of models are proposed.

8.3.1 Lipid – Lipid Bilayer Model

The lipid-lipid bilayer to explain the structure of plasma membrane was given by three scientists namely Overton, Gorion and Grendel. Previously only indirect information was available to explain the structure of plasma membrane. In 1902 Overton observed that substances soluble in lipid could selectively pass through the membranes. On this basis he stated that plasma membrane is composed of a thin layer of lipid. Subsequently then both Gorter and Grendel in 1926 observed that extract from eukaryotic cell membrane was twice the amount expected if a single layer was present. On this basis they stated that plasma membrane is madeup of double layers of lipid maolecules. This model proposed by Gorter and Grendel could not explain the proper structure of plasma membrane but they put the foundation of future models of membrane structure.

8.3.2 Bilayer Model

The lipid bilayer or phospholipid bilayer model is a thin polar membrane made up of two layers of lipid molecule. These membranes are flat sheets forming a continuous barrier around all cells. It is double layer of phospholipids. Outer border of the bilayer membrane is formed by Hydrophilic ends of phospholipids while the inner layer is formed by Hydrophobic layer. The bilayers are particularly permeable to ions which allows cell to regulate salt concentrations and pH by transporting ions across their membranes using proteins called ion pumps. The cell membranes of almost all organisms and many viruses are made of lipid bilayer. Phospholipids with certain head groups can alter the surface chemistry of a bilayer and can serve as signals as well as anchors for other molecules in the membrane of cells. Lipid bilayers are quite fragile and invisible in a traditional microscope. Bilayer creates a "sandwich" style

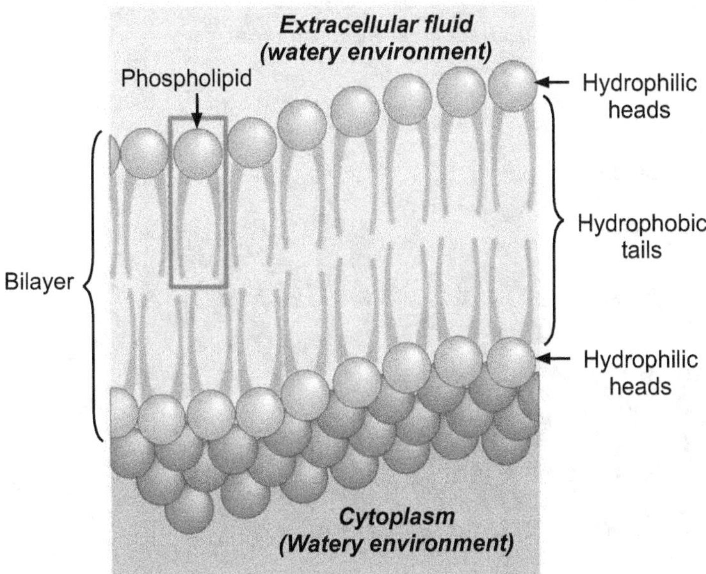

Fig. 8.3: Bilayer Model of Plasma Membrane

arrangement where the hydrophilic heads of each layer face the watery environment inside and outside of the cell. While the hydrophobic tails are confined to the middle creating a hydrophobic region between the two layers of heads. This allows for the plasma membrane to be stable in this dual watery environment. The lipid bilayer is important for maintaining the shape of a cell and for selective permeability. The bilayer is vital for the survival and function of the cell.

8.3.3 Fluid Mosaic Model

This is the universally accepted model for explaining the structure of plasma membrane. It was proposed by Jonathan Singer and Garth Nicolson in 1972. This model postulates that lipid and integrated proteins are disposed in a sort of mosaic pattern and all the biological membranes have **quasi fluid** (half-fluid) structure where both lipid and protein components are able to perform transitional movement within lipid bilayer. In this model, lipid molecules may exhibit intra molecular movement or may rotate about their axis or may display flip-flop movement including transfer from one side of bilayer to the other.

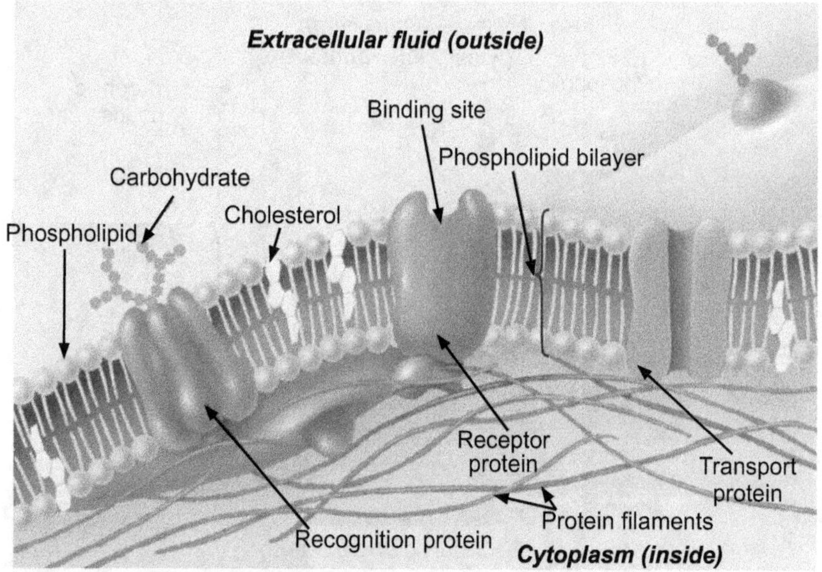

Fig. 8.4: Fluid mosaic model of plasma membrane (Sinner and Nicolson)

Gitler (1972) proposed that in this fluid mosaic model the main components of membrane i.e. lipids proteins and oligosaccharides are held together by means of non covalent interactions. As these molecules have both hydrophilic and hydrophobic properties Hartely (1936) coined the term **amphipathy** to these molecules. Thus lipids and integrated proteins in the plasma membrane are amphipathic in nature.

The important and most recent facts and observations about the fluid mosaic models are:

(i) The intra membrane proteins which are 7-8 nm in diameter traverse the bilayer.

(ii) The membrane proteins are mobile in the plane of this membrane.

(iii) The proteins diffuse laterally in the bilayer but their rate of diffusion is not constant.

This clearly indicates that the lipid bilayer has fluid properties enabling membrane proteins to diffuse rapidly. Even rotational diffuse of proteins is also possible. Afterwards it was noted that not only proteins but lipid molecules also diffuse freely within the lipid bilayers. It was confirmed in biological membranes of mycoplasma, bacteria and red blood cells (RBCs). Basically, the fluid mosaic model proposed by Singer and Nicolson was the modification of Robertson and Davson model. Finally it is confirmed that plasma membrane is a two dimensional oriented solution of integral proteins in the viscous phospholipid bilayer. Recently, many revisions are   made in the original concept of fluid mosaic model proposed to explain the structure of plasma membrane. This model is still relevant to understand the structure and functional dynamics of biological membranes.

8.4 FUNCTIONS OF PLASMA MEMBRANE

1. Primary function of plasma membrane is to protect the cell from it's surroundings.
2. The plasma membrane is selectively permeable to ions and organic molecules and regulate the movement of substances in and out of the cells.
3. It keeps toxic substances out of the cell.

4. It contains receptors and channels that allow specific molecules such as ions and nutrients, wastes and metabolic products that mediate cellular and extra cellular activities.

5. Cell recognition and adhesion: The membrane proteins recognize foreign cells like bacteria and engulf them by phagocytosis. These sites are present on the surface of cell membrane.

6. Antigen specificity: The glycoproteins on the surface of the cell membranes determine the antigen specificities of the cell.

7. Hormone transport: The cell membrane contain receptors which recognize specific hormones required for cellular metabolism, convey the information to the interior of the cell and help this transport.

8. Oxidative phosphorylation: The inner membrane of mitochondria contain the electron transport chain which plays an important role in cellular respiration. During the movement of electrons through this respiratory chain which consists of cytochromes, ATPs are synthesized. This process is known as oxidative phosphorylation.

9. Signal transduction: Plasma membrane receives many intra cellular signals which are transferred to the necessary cells.

POINTS TO REMEMBER

- Plasma membrane is also known as cell membrane, cytoplasmic membrane or plasma lemma.
- Plasma membrane is fluid, phospholipid bilayer that plays a key role in many cellular processes.
- Plasma membrane is selectively permeable.
- It is made up of phospholipids, proteins and carbohydrates like glycoproteins.
- It has many structural and transport proteins as well as many enzymes.
- To explain the structure of plasma membrane two different models are proposed (i) Bilayer model and fluid mosaic model.
- Bilayer model explains that the plasma membrane is made up of two lipid layers. It was proposed by Overton, Gorion and Grendel in 1902 then its more details were given by Gorten and Grendel in 1926.

- The most accepted model for explaination of plasma membrane structure was proposed by Singer and Nicolson in 1972.
- It showed that in plasma membrane lipid and proteins are disposed in a sort of mosaic model.
- This model proposed that plasma membrane has fluidity hence the name of the model is "Fluid Mosaic model".
- According to this model plasma membrane is made up of lipids, proteins and oligosaccharides held together by non-covalent bonds (Gitler, 1972).
- Both the lipids and proteins in plasma membrane are amphipathic in nature.
- Plasma membrane performs many important functions like: (i) Protection of cell (ii) Signal transduction, (iii) Controlling the in and out movement of various ions and organic molecules in the cells as it is selectively permeable (iv) keeping toxic substances out of the cells (v) regulate transport of hormones and other molecules (vi) recognization and adhesion (vii) antigen specificity, oxidative phosphorylation etc.

EXERCISE

1. Define plasma membrane and give its composition.
2. Explain the general structure of plasma membrane.
3. Give an account of various components of plasma membrane and add a note on their location in plasma membrane.
4. Explain the "lipid bilayer model" of plasma membrane with suitable diagram.
5. Explain "Fluid mosaic model" of plasma membrane with suitable diagram.
6. Compare: Bilayer and Fluid mosaic model.
7. Explain why "Fluid mosaic model is universally accepted.
8. Give an account of various functions performed by plasma membrane.
9. Write notes on:
    (i) Plasma membrane structure
    (ii) Components of plasma membrane

    (iii) Bilayer model

    (iv) Fluid mosaic model

    (v) Functions of plasma membrane

    (vi) Signalling and signal transduction by plasma membrane

    (vii) Proteins in plasma membrane

    (viii) Phospholipids in plasma membrane

10. Sketch and label the structure of :

    (a) Phospholipid bilayer

    (b) Fluid mosaic model

    (c) Plasma membrane

11. Explain why "both the lipids and proteins in plasma membrane are amphipathic in nature".

12. Explain: what is mean by *Hydrophilic* and *Hydrophobic* molecules.

❖❖❖

9

CHAPTER

STRUCTURE AND FUNCTIONS OF CHLOROPLAST, MITOCHONDRIA AND ENDOPLASMIC RETICULUM

◆ POINTS TO LEARN ◆

9.1   **Introduction**

9.2   **Ultrastructure and Functions**

     9.2.1 **Chloroplast**

     9.2.2 **Mitochondria**

     9.2.3 **Endoplasmic reticulum**

*   **Points to Remember**

*   **Exercise**

9.1 INTRODUCTION

All the plant cells have various types of cell organelles suspended in the cytoplasm e.g. chloroplast, mitochondria, endoplasmic reticulum, Golgi apparatus, lysosomes, ribosomes, nucleus etc. These cell organelles are actively engaged in various types of activities. Each cell organelle has definite structure and a very specific role to play in the cell. We will discuss few of them.

9.2 ULTRASTRUCTURE AND FUNCTIONS

9.2.1 Chloroplast

Chloroplasts are the master molecules in the process of photosynthesis. They harvest the solar energy and convert it into useful biological or chemical energy. Chloroplasts are present only in green plants, algae and in some bacteria. They are ultramicroscopic structures, differ in size and shape. The chloroplast of higher plants are disc shaped or oval in structure about 10 μm in length and

2-4 μm in diameter. Algal species contain larger chloroplast and differ in shapes. e.g. spiral in *Spirogyra* cup shaped in *Chlorella*. The number of chloroplast, is 1-2 per cell in algae, while in higher plants the number is very high.

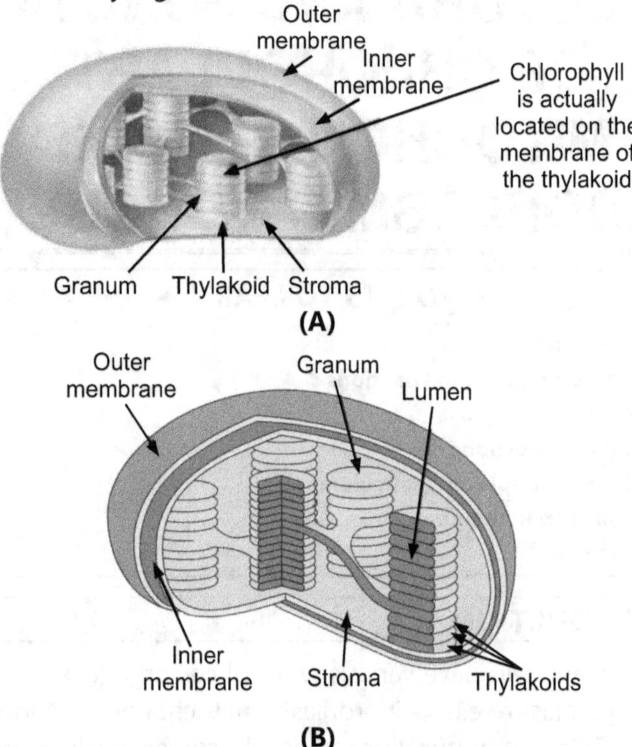

Fig. 9.1: Ultrastrcutre of Chloroplast A and B

The chloroplast is a double membranous structure i.e. the outer membrane and inner membrane. The space between the two membranes is known as inter membrane space.

Outer membrane: It is semi porous and permeable to small molecules and ions, which diffuse easily.

Inner membrane: It forms a border to stroma. It regulates passage of molecules in and out of chloroplast.

Inter membrane space: It is present in between the two membranes which is 10-20 nanometer.

Chloroplast shows matrix or stroma in which grana are present. A single chloroplast may show presence of 50 or more grana.

Granum: Each granum is made up of stacks of thylakoids within these, space is found which is known as thylakoid space. Each granum looks like a stack or pile of coins, measuring 0.5 to 2 μm in diameter, light reaction of photosynthesis takes place in grana.

Stroma: It is an alkaline aqueous fluid or amorphous matrix, which is protein rich, distributed throughout the chloroplast. The chloroplast DNA, ribosomes and the thylakoid system, starch granules, many proteins and different enzymes of Calvin cycle of photosynthesis are found floating in stroma.

Thylakoid system: It is a membranous sacks suspended in stroma. The chlorophyll-green pigment is found in thylakoid. The thylakoids are arranged in stacks known as grana. Each granum contains 10-20 thylakoids.

Grana and Stroma lamellae: The lamellae which connect grana are known as grana lamellae while the lamellae present in stroma are called as stroma lamellae.

Functions of Chloroplast

1. The main function of chloroplast is to synthesize carbohydrates during the process of photosynthesis from CO_2 and H_2O using light energy:

$$6\ CO_2 + 12\ H_2O \xrightarrow[\text{Chlorophyll}]{\text{Light}} C_6H_{12}O_6 + 6H_2O + 6O_2 \uparrow$$

 This reaction takes place in stroma.

2. Chloroplast in the process of photosynthesis absorb CO_2 from the atmosphere and help to reduce air pollution.

3. During the process of photosynthesis which takes place in chloroplast oxygen is released in the atmosphere, thus maintaining the concentration of O_2 in air. It takes actually takes place in grana.

4. Chloroplasts are the main molecules harvesting light energy (sunlight) and converting it into useful biological (chemical) energy.

5. During the process of photosynthesis ATP (energy molecule) and Glucose ($C_6H_{12}O_6$) molecules are formed along with NADPH (high

energy molecule). These reactions take place in PS I and PS II present in granum.

6. The green colour of plant leaves and some algal members is due to chlorophyll pigment.

7. Photolysis of water (H_2O) into presence of sunlight forming H^+(Hydrogen ions) and OH^- ions, which combines releasing oxygen in the atmosphere.

    (i) $H_2O \longrightarrow H + OH$

    (ii) $OH + OH \longrightarrow O_2H - H_2$

8. Photophosphorylation takes place in grana by addition of ip inorganic phosphate to ADP – giving rise to ATP and NADH

$$ADP + ip \xrightarrow[\text{Light}]{\text{Grana}} ATP + NADH$$

The photophosphorylation reactions are of two types:
(i) Cyclic and (ii) Noncyclic.

9.2.2 Structure of Mitochondria

Mito-thread, chondrion-granule. This term was coined by Benda (1898). Mitochondria were first observed by Richard Altman (1894). Mitochondria are the most important cell organelles in all the plant and animal cells except some bacteria and fungi. These are popularly known as **"Power house of the cell"**. They are closely associated with aerobic respiration, where Glucose molecules are broken down to CO_2 and H_2O in presence of oxygen, releasing energy in the form of ATP.

Mitochondria usually found in cytoplasm which are small granular or filamentous bodies, even they vary in size, shape and number from cell to cell. The average length of mitochondria is 3-5 μ and the average diameter is 0.5 to 1 μ. The number of mitochondria present in each cell depends on the metabolic activities in the cell. The main structure of mitochondria consist of five different parts:

(1) Outer mitochondrial membrane
(2) The inter membrane space
(3) The inner mitochondrial membrane
(4) The cristae
(5) The matrix the space within the inner membrane.

Inner membrane
Outer membrane
Lumen
Stroma lamellae
Thylakoid
Intermembrane space
Granum
Stroma

(A)

Intermembrane space
Ribosome
Granule
Matrix
ATP synthase particles
DNA
Outer membrane
Inner membrane
Cristae
0.1 ~ 0.5 µm
1 ~ 2 µm

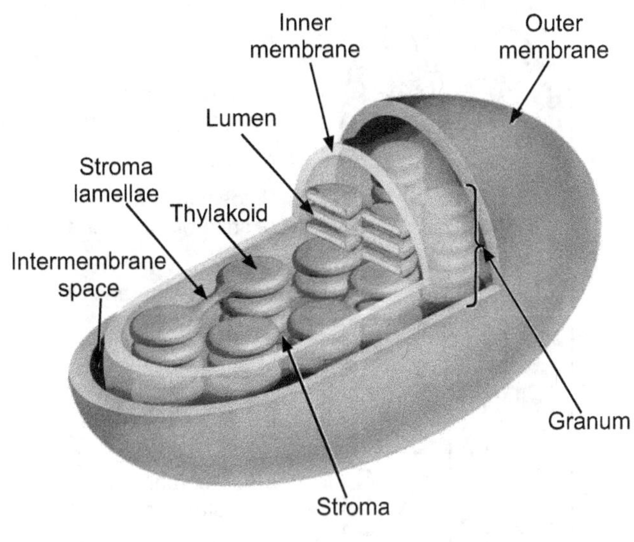

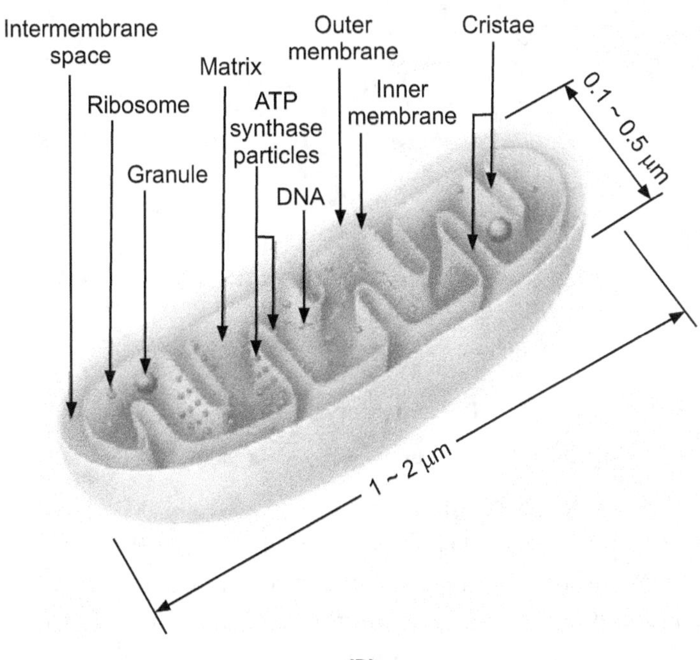

(B)

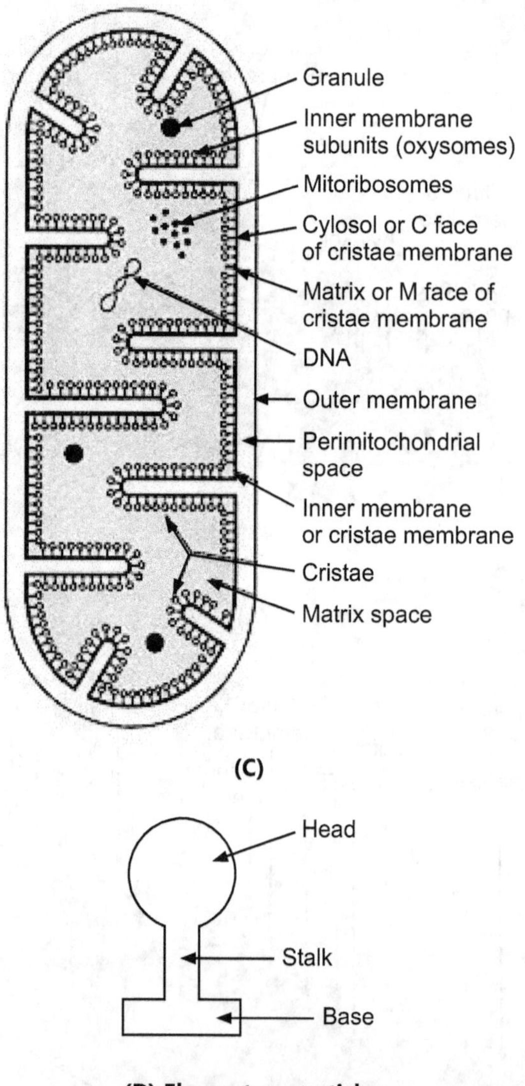

(D) Elementary particle

Fig. 9.2: Structure of Mitochondria (A, B, C & D)

9.2.2.1 Outer Membrane

It is made up phospholipid bilayer and 60-75A° thick, containing integral proteins known as porins. Its major trafficking part is "VDAC", i.e. voltage dependent anion channel. It is primary transporter of nucleotides, ions and metabolites between the cytosol and the inter membrane space. The outer membrane also contain many enzymes. The outer membrane constitutes 50% of mitochondria.

9.2.2.2 Inter Membrane Space

It is the space between two membranes which is known as "peri mitochondrial space". It is about 40-70 A° in width, filled with watery fluid. It has high proton concentration.

9.2.2.3 Inner Mitochondrial Membrane

The inner membrane of mitochondria is folded into cristae. The membrane is permeable only to oxygen, CO_2 and H_2O. The inner mitochondrial membrane contain proteins, that perform redox reactions in oxidative phosphorylation. On this membrane ATP synthesis enzyme, transport proteins, and several antiport systems are present.

9.2.2.4 Cristae

These are the folds of inner mitochondrial membrane for expanding its surface area for enhancing its ability to produce more and more ATP molecules. It contains about $\frac{2}{3}$ of total proteins present in mitochondria. These cristae produce mainly ATP molecules with the help of ATP synthase enzyme. They also show presence of t-RNA, mitochondrial DNA genome. The cristae are having thousands of small particles on them, which are stalked, these particles are with base piece, a stalk and head piece. These are spaced at a distance of 100 A°. The head piece is 75-100 A° in diameter and the stalk is about 50A° in length. These particles (elementary particles) are involved in electron transport and oxydative phosphorylation. These are also known as ETP (electron transfer particles). Recently these are known as F_1 particles.

9.2.2.5 Mitochondiral Matrix

Mitochondria possess a gel like material known as mitochondrial matrix, which is highly viscous. All the Citiric acid cycle (Kerbs cycle) enzymes are present in matrix. The matrix is filled with water and proteins and many enzymes and mitochondrial DNA, ribosomes, small organic molecules, nucleotides, cofactors and inorganic ions.

9.2.2.6 Functions of Mitochondria

1. **Power house of a cell:** The prominent function of mitochondria is to produce ATP the cellular energy for all metabolic activities because of this mitochondria are known as power house of cell.

2. **Cellular respiration:** Mitochondria are the site of all types of cellular respirations e.g. Glycolysis, oxidation of pyruvic acid, Krebs/ Citric acid cycle and oxidative phosphorylation during which complex glucose molecules are oxidized to CO_2 + H_2O releasing huge amount of energy in the form of ATP, $NADH_2$ and $FADH_2$.

3. **ETS:** Respiratory chain: The cristae on inner membrane of mitochondria and F_1 particles are having the respiratory chain also known as electron transport systems (ETS) which is engaged in oxidative phosphorylation. It consist of cyt a, cyt b, cyt c etc. The ATPs are synthesized during the transport of electrons through this chain.

4　**Transamination:** In mitochondria, many times α, keoglutarate is converted to glutamate, proline and arginine during the processes of transamination. Similarly oxalo acetic acid (OAA) is converted to asperate and aspergine during the same process. All the amino acids produced are used by the cell in different processes as and when required.

5. **Protein synthesis:** The mitochondria have their own set of DNA which is used to synthesize proteins.

6. **Storage of calcium ion:** In many plant and animal cells mitochondria help in storage of calcium.

7. **Site of several metabolic reactions:** It is playing very active role in aerobric respiration, protein synthesis, transamination and transport process.

8. **Beta oxidation:** Many fatty acids are synthesized in plant cell through β-oxidation.

9. **Production of heat:** During aerobic respiration glucose molecules are oxidized along with production of heat. Lot of heat.

10. **Signaling:** Mitochondria through their activities play major role in transfer of different types of signals between cells and different organelles especially through reactive oxygen species (ROS).

11. **Regulation of cellular metabolism:** Almost all the cellular/metabolic activities in the cells are regulated or controlled by mitochondria. They play a vital role in the regulation of metabolic activities of a cell.

12. **Other function:** In addition to above functions, mitochondria play very important role in (a) Apoptosis: i.e. programmed cell death, (b) Heme synthesis reactions, (c) Steroid synthesis, (d) Hormonal signaling.

9.2.3 Endoplasmic Reticulum (ER)

ER was first observed by Garnier in 1897. The term ER means network of fabric of membranes. Endoplasmic reticulum (ER) like chloroplast and mitochondria is very important cell organelle present in the eukaryotic cells of plants as well as animals, however it is absent in prokaryotes. It is a large dynamic structure consist of cisternae, tubules and vesicles. ERs are distributed throughout the cytoplasm and on an average it occupies 10% of cell volume.

The Cisternae:

These are broad, flattened sacs of membrane bond structures and about 40-50 A° thick and often arranged in parallel stacks and inter connected with other.

Tubules:

These are of more diverse nature in shapes then cisternae. Their diameter ranges from 50 to 100 mm.

Vesicles:

These are rounded in shape and their diameter ranges from 20 to 500 μm. The membranes of ER are continuous with the outer nuclear membrane.

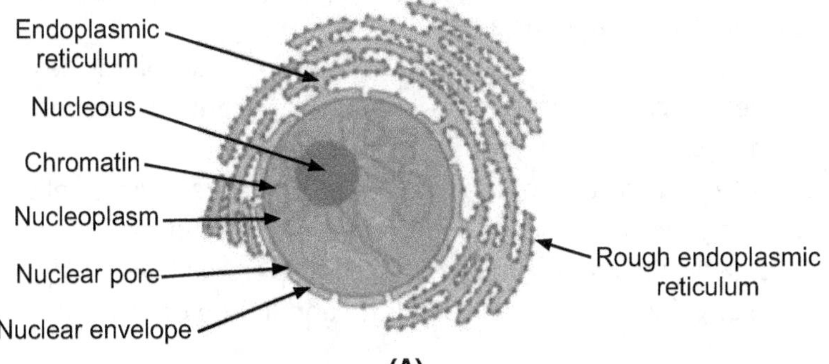

(A)

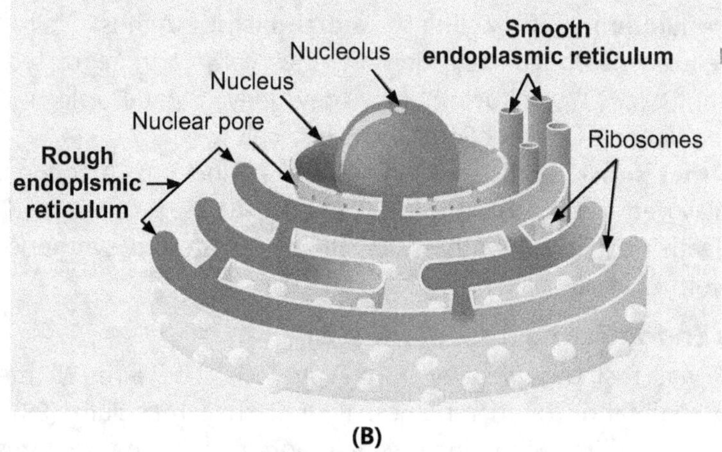

(B)

Fig. 9.3: Structure of endoplasmic reticulum (A & B)

Types of Endoplasmic reticulum:

There are two main types of ER:

(i) Smooth endoplasmic reticulum (SER)

(ii) Rough endoplasmic reticulum (RER)

(i) Smooth endoplasmic reticulum (SER):

It is a granular form of ER. It is called as smooth ER because its membranes are not covered with ribosomes. Smooth ER generally very common in cells which synthesize nonprotein substances like phospholipid, glycolipid and steroidal hormones. SER is less stable than RER.

Functions:

1. Smooth ER carries out many functions other than protein synthesis.
2. Smooth ER is synthesizing lipids, phospholipids and steroids.
3. It acts a storage organelle.
4. It also carries the metabolism of carbohydrates and detoxification of natural metabolic products, alcohol and drugs.
5. It regulates calcium ion concentration in the muscle cells.
6. The process of Gluconeogenesis is carried out by SER.
7. It serves as transitional area for transport vehicles.
8. It also functions in carbohydrate and lipid biosynthesis.

Both the SER and RER are typically connected to each other so that the proteins and membranes made by RER can freely move into the SER for transport to other parts of the cell.

(ii) Rough endoplasmic reticulum (RER):

Structure:

Rough ER is also known as granular ER because its membranes are covered with full of ribosomes. RER are extensively formed in the cells which are actively synthesizing proteins and it is more stable one. The ribosome on RER are constantly being bound and released from its surface. The presence of large rough ER is indicative of metabolically active cell. The binding site of ribosomes on rough ER is known as translocon. The rough ER is located near and is continuous with the outer layer of the nuclear membrane. The RER and other cell organelles are held closely together allowing the transfer of lipids and other small molecules.

Functions:

1. RER is mainly concerned with protein synthesis.
2. The synthesized proteins are transported to golgi apparatus.
3. It plays an important role in the synthesis of different types of membranes.
4. RER produces many types of transmembrane proteins and lipids for the synthesis of membranes of cell organelles like lysosomes, secretary vesicles, Golgi appratus and cell vacuoles.
5. It synthesize the membrane and secretary proteins.
6. The RER in WBCs produces antibodies.
7. In pancreatic cells RER produces insulin.

POINTS TO REMEMBER

- Eukaryotic cells in their cytoplasm contain many important cell organelles like chloroplast, mitochondira, endoplasmic reticulum, Golgi complex etc.

- The cell organelles are performing specialized functions.

- The most-important cell organelle present in the plant cell and some algae is chloroplast, containing green colour pigment chlorophyll.

- Chloroplasts are seen only under EM.
- They have double membranous envelop enclosing matrix (stroma) and grana.
- Its major function is to perform photosynthesis and produce carbohydrates, ATP and the release of oxygen.
- Chloroplast harvest (trap) solar energy and convert it into useful biological/chemical energy.
- During the process of photosynthesis atmospheric CO_2 combines with H_2O and produce glucose. Thus helping in carbon sequestration.
- Light reaction of photosynthesis take place in granum while carbon reduction (Calvin cycle) occur in stroma.
- Mitochondria is the most important cell organelle in plant and animal cell. It is also a double membranous structure.
- The outer membrane is smooth while the inner membrane is folded, these folds are known as cristae.
- Mitochondria also contain a gel like viscous fluid which is known as matrix.
- Inter membrane space is present between the two membranes.
- F_1 particles are present on cristae and they are involved in synthesis of ATP through oxidative phosphorylation.
- F_0-F_1 particles or ATP synthase enzyme is actually involved in ATP synthesis.
- Mitochondria is known as "Power house" of a cell.
- Mitochondria is the place of aerobic respiration in which glucose molecule is broken down to CO_2 + H_2O releasing energy (ATPs) and heat.
- The different steps of respiration e.g. Glycolysis, Kerbs cycle and terminal oxidation occur in mitochondria.
- ETS (respiratory chain) is found in F_1 particles.
- Transamination, beta oxidation, protein synthesis and storage of calcium ions are some additional functions performed by mitochondria.

- Mitochondria also plays an important role in signal transduction, regulation of cellular metabolism, and apoptosis.
- Endoplasmic reticulum (ER) is a large dynamic structure consisting of cisternae, tubules, and vesicles.
- There are two main types of Endoplasmic reticulum: (a) Smooth ER and (b) Rough ER.
- SER is also known as agranular ER. It is known as smooth because it has no ribosome attached.
- SER is very common in cells which are synthesizing non-protein substances like phospholipid, glycolipid and steroidal hormones.
- SER is synthesizing lipids, phospholipids and steroids.
- Rough endoplasmic reticulum is granular and having a lot of ribosomes attached to it looking rough hence the name.
- It is more stable than SER.
- RER is found near and continuous with nuclear envelop.
- Main function of RER is protein synthesis with the help of ribosomes attached to it.
- It also take part in synthesis of different types of membranes of cell organelles.

EXERCISE

1. Make a list of different types of cell organelles found in eukaryotic plant cell and briefly explain the structure of chloroplast.

2. Sketch and label the ultramicroscopic structure of chloroplast and add a note on its function.

3. With the help of well labelled diagram explain the functions of mitochondria.

4. Give an account of the structure of endoplasmic reticulum.

5. Explain why mitochondria is known as the "Power house of cell".

6. Justify: Chloroplasts are the master molecules in harvesting of sunlight.

7. Comment on: Smooth and rough endoplasmic reticulum.

8. Write briefly on:
    (a) Structure of chloroplast
    (b) Structure of mitochondria
    (c) Structure of endoplasmic reticulum
9. Explain the function of stroma and grana in photosynthesis.
10. Explain the role of mitochondria in aerobic respiration.
11. Describe the role of smooth and rough endoplasmic reticulum in synthesis of proteins and lipids.
12. Write short notes on: Cell organelles studied by you.
13. Give the site of synthesis of:
    (a) ATPs
    (b) Break down of glucose
    (c) Light reaction of photosysntesis.
    (d) Dark reaction of photosynthesis.
    (e) Protein synthesis.
    (f) Oxydative phosphorlyation.
    (g) Photophosphorylation.
14. Distinguish between:
    (1) Stroma and grana
    (2) Smooth and rough endoplasmic reticulum
    (3) Outer and inner membrane of mitochondria
    (4) Photophosphorylation and oxidative phosphoralytion

10

CHAPTER

CELL CYCLE IN PLANTS

◆ POINTS TO LEARN ◆

10.1 Introduction

10.2 Stages/Phases of Cell Cycle

    10.2.1 Interphase

    10.2.2 G_1-Phase or Post Mitotic Phase

    10.2.3 G_2-Phase

    10.2.4 S-Phase

    10.2.5 M-Phase

10.3 Importance of Cell Cycle in Plants

10.4 Divisional Stages of Mitosis and Meiosis

    10.4.1 Divisional Stages of Mitosis

    10.4.2 Meiotic Type of Cell Division

***   Points to Remember**

***   Exercises**

10.1 INTRODUCTION

Cell cycle can be defined as the "Life cycle" of a cell. The cell cycle is passing through series of precisely timed and carefully regulated stages of growth, DNA replication and division.

The cell cycle in plants and animals is very specific and an ordered set of events, leading to cell growth and its division in to two daughter cells. The cell cycle begins with the formation of daughter cells at the end of telophase. The daughter cells are smaller than the parent cell and their DNA content is half of the parent cell. These daughter cells grow in size by synthesizing new cytoplasmic and nuclear materials till the total volume become four times of the original volume and the DNA content get doubled by its replication. After this the fully developed cells are again ready to under go division. In short cell cycle is the series of growth and development steps, which cell under goes between its birth and reproduction.

(10.1)

10.2 STAGES/PHASES OF CELL CYCLE

For the division of cell it must complete several important steps. e.g. it must grow, copy its genetic material (DNA) and physically divide into two daughter cells. The cells perform these tasks in an organized, predictable series of steps that make up the cell cycle. It is called as cell cycle because at the end of each go-round the two daughter cells can start, the exactly same process once again from the beginning.

The cell cycle is divided into two major phases:

(1) Interphase or non-dividing phase

(2) The mitotic phase (karyokinesis) or dividing phase. It is also known as M-phase.

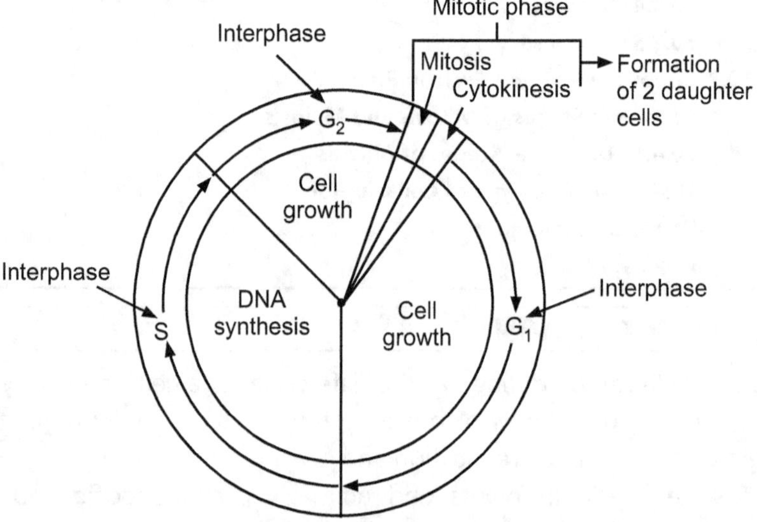

Fig. 10.1: Stages of Cell Cycle

10.2.1 Interphase

During this phase, the cell grows and makes a copy of its own DNA. It is the preparation by the cell itself for further divisions.

Phases of cell cycle: The scientist Howard and Pele (1953) have divided cell cycle into four phases G_1, S, G_2 and M phase.

The G_1 and G_2 are known as Gap phases, M phase is of mitosis and S phase is for DNA synthesis. The G_1, S and G_2 phases are combined to form classical Interphase. During this phase the cell is fully preparing it self for next mitosis phase. This not a resting phase, but many preparatory activities are going on in this phase. Most of the cells usually spend their maximum life span in the interphase.

10.2.2 G_1-Phase or Post-Mitotic Phase

The daughter cell enters into G_1 phase of interphase of the new cell cycle. G_1-Phase, the resting phase is also called as the first gap phase; because during this phase there is no DNA synthesis. Recently G_1 has been termed as Fist growth phase since it involves synthesis of RNA, proteins and membranes which lead to the growth of nucleus and cytoplasm of each daughter cell towards the mature size.

During this phase, the young daughter cell grows in size by synthesizing cytoplasm. It also synthesize many enzymes mRNA, tRNA, ribosomes, and proteins and stores them. It also start synthesizing the nitrogen bases required for DNA replication. During this phase, chromosomes are extended to form slender fine threads and form an interwoven network of chromatin.

The main function of the G_1 phase is to prepare nuclei for DNA synthesis in the S phase. In G_1 phase the cell accomplishes the majority of its growth along with the synthesis of mRNA and proteins required in subsequent steps. The G_1 phase also acts as crucial checkpoint allowing cells to decide whether they can truely commit to mitotic division. In plants, cells in G_1 phase are described as quiescent. The control between cell quiescence and proliferation allows them to respond to factors such as nutritional availability and abiotic/biotic stress. During G_1 phase, which is also known as "first gap phase the cell grows physically larger, make the several copies of different organelles" and makes the molecular building blocks required for further steps. G_1 is the longest phase of cell cycle and cells usually remain for 10 hrs in G_1 phase of 24 hrs of the total time of cell cycle.

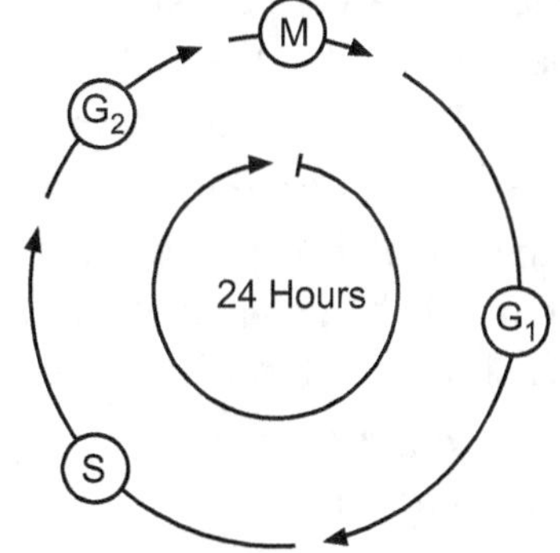

Fig. 10.2: Time Duration in Phases of Cell Cycle

10.2.3 G_2-Phase

It is the second gap phase. During this phase cell grows more and more, makes (synthesize) proteins and organelles and begins to re-organize its contents for the preparation of mitosis, G_2 phase ends when mitosis begins. During this stage r-RNA, m-RNA and nuclear RNA are synthesized. Its the intermediate step between the completion of DNA synthesis and the beginning of DNA segregation. It is the post DNA synthesis period, during which the cell makes preparations for the initiation of mitosis. G_2 is the shorter phase casting for only 3-4 hrs in most of cells.

10.2.4 S-Phase

It is the phase of complete synthesis of DNA in its nucleus. In this phase, there is also duplication of centrosome, which helps the separation of DNA during M-phase. Usually cells take 5-6 hrs to complete S-phase. It is very essential phase of cell cycle. Once DNA replication is completed, the cell has twice the number of chromosomes and become ready to enter the G_2 phase.

The G_1, S and G_2 phases together constitute the interphase.

10.2.5 M- Phase

During the mitotic (M) phase the cell divides its copied DNA and cytoplasm to make two new cells. M-phase involves two distinct divisions related processes: (i) Mitosis, (ii) Cytokinesis. In mitosis the Nuclear DNA of the cell condenses into visible chromosomes and is pulled a part by the mitotic spindlier. Mitosis takes place in four stages, (i) Prophase, (ii) Metaphase, (iii) Anaphase and (iv) Telophase.

In cytokinesis the cytoplasm of a cell divides into two by the formation of cell plate and finally giving two new daughter cells. Cytokinesis begins just at the end of mitosis. In most of the cells M-phase is completed with in 2 hrs.

The duration of cell cycle varies from organism to organism and from cell to cell.

In short the cell cycle is a four stages process in which cell increases in size (G_1) copies its DNA (S.phase) prepare to divide (G_2) and finally it divides (M-phase).

10.3 IMPORTANCE OF CELL CYCLE IN PLANTS

(1) It plays a crucial role in a living organism's life cycle.

(2) In case of multicellular organism it helps in growth and repair by producing large number of identical cells.

(3) Almost all plants and animals depend on the cell cycle for their growth and development.

(4) The cell cycle is the main process of regeneration and repair in plants and animals.

(5) It helps in equal distribution and making the constant number of chromosomes in all the cells of bodies of plants and animals.

10.4 DIVISIONAL STAGES OF MITOSIS AND MEIOSIS

10.4.1 Divisional Stages of Mitosis

Mitosis or somatic cell division is the multiplication of somatic cells or body cells of plants into daughter cells of equal size both containing the same number of chromosomes as the parent cell. The term mitosis Greek word refers to thread like appearance of chromosomes in the early cell division. During mitosis the nucleus gets completely reorganized. The cell enters in mitosis it starts with Interphase mitotic cell division takes place through series of consecutive phases such as

(1) Prophase

(2) Prometaphase

(3) Metaphase

(4) Anaphases

(5) Telophase

(6) Cytokinesis.

In mitosis series of changes take place in nucleus and cytoplasm. The changes in nucleus are known as Karyokinesis while the changes in cytoplasm are known as cytokinesis.

(1) Prophase: It is subdivided into (i) Early prophase, (ii) Middle prophase, (iii) Late prophase (prometaphase)

It is the largest stage of mitosis and may lasts for several hours.

(a) Early Prophase: The cell starts to break down some structures and build others. Setting the phase for division of chromosomes. The chromosomes start to condense (making them easier to pull apart later on). The mitosis spindle begins to form. The nucleolus disappears. This is the sign that the nucleus is getting ready to breakdown (Fig. 10.4 A).

(b) Late Prophase: It is also known as prometaphase. The mitotic spindle begins to capture and organise the chromosomes. The chromosomes finish condensing to they are very compact. The nuclear envelop breaks down, releasing the chromosomes. The mitotic spindle grows more and some of the microtubules starts to capture chromosomes (Fig. 10.4 B).

Mitotubules can bind the chromosomes at kinetochore. Centromeres are the regions of DNA, where the sister chromatids are mostly tightly connected.

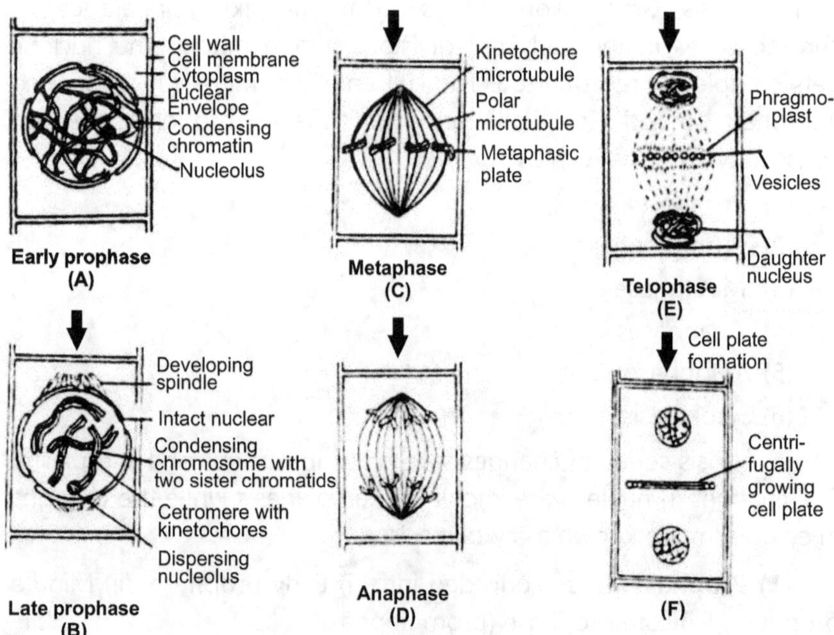

Fig. 10.3

Metaphase: The spindle has captured all the chromosomes and lined them up at the middle of the cell ready to divide. All the chromosomes align at the metaphase plate. The two kinetochores of each chromosome are attached to microtubules from opposite spindle poles. Before proceeding to anaphase the cell check to make sure that all the chromosomes are at the metaphase plate with their kinetochroes correctly attached to microtubules. This is called as Spindle check point (Fig. 10.4 C).

Anaphase: In anaphase the sister chromatids separate from each other and are pulled towards opposite poles of the cell. All this is done by motor proteins. The anaphase is further subdivided into early anaphase, middle anaphase and late anaphase depending on the movement of chromosomes towards the two opposite poles (Fig. 10.4 D).

Telophase: It is also subdivided into early telophase and late telophase. The cell division is nearly completed and it starts to re-establish its normal structures as cytokinesis takes place.

The mitotic spindle is broken down into its building blocks. Two new nuclei are formed one for each set of chromoses. Nuclear membranes and nucleoli reappear. The chromosomes begin to decondense and return to their 'stringy' form.

Thus, telophase is the reverse of prophase. The spindle fibres disappear (Fig. 10.4).

Cytokinesis: The division of the cytoplasm to form two new daughter cells. The cell plate is formed during cytokinesis at the middle of cell, separating the two daughter cells.

Significance of Mitosis:

Mitosis is very important for plants and animals because it is responsible for their growth and development. As well as it is helping in repair and damage recover of organisms. It also plays important role in regeneration of plant parts. Through mitosis equally constant number of chromosomes in each cell is maintained. It takes place in somatic cells body cells of plants and animals (Fig. 10.4).

10.4.2 Meiotic Type of Cell Division

Meiosis: This type of all division takes place only in reproductive cells. It is pivotal in plants and animals for reducing the chromosome number of diploid organisation (2n) to half (n).

Genetic resortment takes place during meiosis. Meiotic recombination and chrosomosome segration takes place through meiotic type of cell division. Meiosis is a specialized cell division that generate four haploid daughter cells from a diploid parent cell, in which single round of DNA, replication and two consecutive rounds of nuclear division takes place. It's a reduction cell division.

Following are the different stages of meiosis in plants:

(I)   First Meiotic Cell Division:

    (A)   Prophase I: It consists of following stages

    (i)   Leptotene

    (ii)   Zygotene

    (iii) Pachytene

    (iv)   Diplotene

    (v)   Diakinesis

    (B)   Metaphase I including prometaphase

    (C)   Anaphase – 1

    (D) Telophase – 1

(II)   Second mitotic cell division:

    (i)   Interphase or interkinesis

    (ii)   Prophase II

    (iii) Anaphase II

    (iv) Telophase II

All the stages are summarised in the chart.

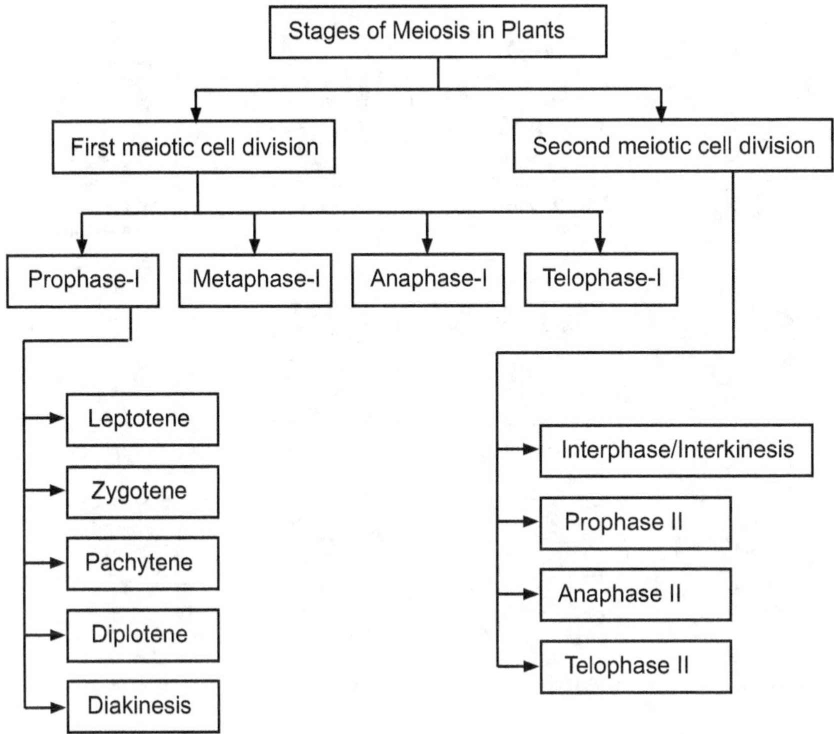

- At the end of first meiotic cell division two daughter cells are formed.

- Its reduction division.

- Diploid mother cell forms two haploid daughter cells.

- At the end of second meiotic cell division four daughter cells are formed.

- Its equitional division

- The two daughter cells just under go divisions like mitosis and produce four haploid cells.

(I) Prophase I: Consists of five sequential cell divisions (i) Leptotene, (ii) Zygotene, (iii) Pachytene, (iv) Diplotene, (v) Diakinesis.

(i) Leptotene: It is the first stage of prophase I (meaning thin threads). Individual chromosomes begin to condense into long strands within the nucleus. Nucleus increases in size. The synthesis of RNA and protein increases. (Fig. 10.5 A)

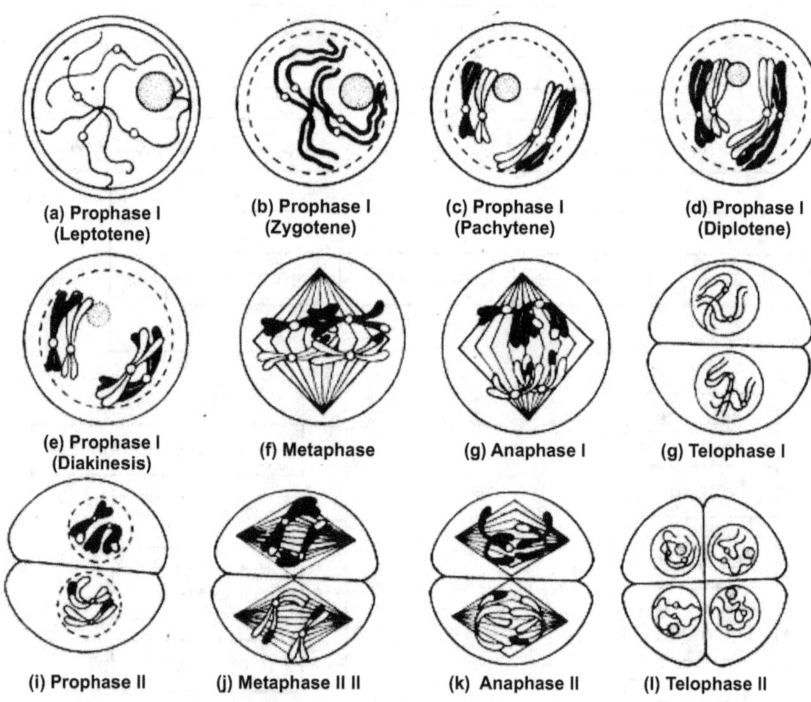

Fig. 10.5: Stages of Meiosis

It is also known as Leptonema.

(ii) Zygotene: This is also known as zygonema. The name zygotene to this stage indicate that the chromosomes form paired thread like structure. The chromosomes become shorter and thicker. The pairing or **synapsis** of homologous chromose take place. The homologus chromosomes undergo length wise pairing in which one chromosomes is from paternal and the next is from **maternal** side. This stage is known as **'bivalent'** stage. Each chromosome consists of four chromatids and hence known as **'tetrad'**. The pairing of chromosomes is very specific forming a special structure called as **Synaptonemal complese**. It is the structural basis for pairing and synopsis of meiotic chromosomes (Fig. 10.5 B).

(iii) Pachytene: It is also called as pachynema meaning thick threads. The homologus chromosomes now are very much closely associated than in the zygotene. This process is known as **synapses**.

The chromosomes are in bivalent or tetrad condition. The synapsed chromosomes generally undergo the process of **crossing over** in pachytene stage. The chromosomes continuously show condensing. Each chromatid is the unit of crossing over (Fig. 10.5 C).

(iv) Diplotene or Diplonema: Meaning two threads. The homologus chromosomes begin to separate. The chromosomes are held at one or more points known as **'chaismata'**. The crossing over or recombination takes place at chaismata. At least one chaisma is formed for each bivalent. In this stage the four chromatids become visible and the **'Synaptonemal complex'** disappear (Fig. 10.5 D).

(v) Diakinensis: Chromosomes condense further in this stage, diakinesis meaning 'moving through'. In diakinesis the four parts of the tetrads are actually visible. The homologous chromosomes separate further and the chiasmata terminalise, making them clearly visible (Fig. 10.5 E).

(II) Metaphase I: The chromosomes are arranged at the equator plate of the cell. The spindle is formed which is attached to the centromeres of two homologous chromosomes. The centromers of each bivalent lie on opposite sides of equatorial plate. The bivalent consisting of two centromeres are being pulled towards the poles of opposite side (Fig. 10.5 F).

(III) Anaphase I: Two homologous chromosomes of each bivalent separate and start moving towards the opposite poles of the cell. The short chromosomes separate rapidly but long chromosomes take time for the separation. In each homoglous chromosome one chromatid is unchanged while the other has undergone mixing of maternal and paternal section (Fig. 10.5 G).

(IV) Telophase I: It begins when the chromosomes reach to their respective poles. The chromosomes for some time remain in condensed state. Later they under go depolarization and become elongated. Nuclear membrane is formed around each set of chrosomes. By the formation of cell plate between two groups of chromosomes two daughter cells are formed due to cytokinesis (Fig. 10.5 H).

Second Meiotic Division (Meiosis II):

During this stage actually mitotic divisions occur in two daughter cells formed during meiosis I and giving rise to four haploid daughter cells. It consists of

(1)   Interphase II or Interkinesis

(2)   Prophase II

(3)   Metaphase II

(4)   Anaphase II

(5)   Telophase II

(6)   Cytokinesis

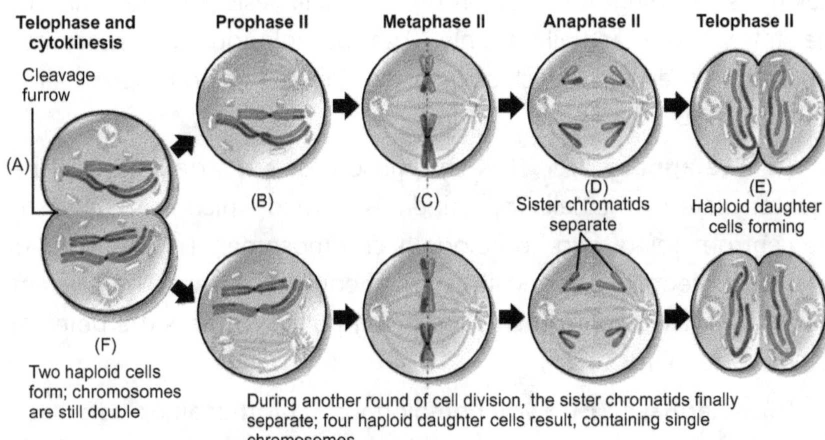

Fig. 10.6

Interphase II or Interkinesis:

The two haploid daughter cells undergo typical resting stage as in mitosis. This intervening stage between telophase I and beginning of prophase II is called as Interphase II or Interkinesis in which there is no DNA duplication (Fig. 10.6 A).

Prophase II: This is a very short duration stage or very simple. The chromosomes undergo shortening and thickening. The sister chromatids have already been separated. The two chromatids of each chromosome remain separate except at centromere. Spindle formation takes place in prophase II along with the disappearance of nuclear membrane (Fig. 10.6 B).

Metaphase II: The chromosomes become arranged on the metaphase plate. The centromere of each chromosome splits longitudinally forming two centromeres. At the end of metaphase II centromeres are connected to the spindle fibres from respective poles (Fig. 10.6 C).

Anaphase II: It begins with the movement of chromatids to the opposite poles and they are pulled by contraction of spindle fibres. The sister chromaids are now become individual chromosomes reach towards the opposite poles of the cell (Fig. 10.6 D).

Telophase II: Nuclear membrane is formed around each set of chromosomes, the nucleolus in each nucleus. The chromosomes lengthen and become indistinct (Fig. 10.6 E).

Cytokinesis: Cell walls are formed between the daughter cells at the end of Meiosis II. Finally, four daughter cells with haploid number of chromosomes are formed (Fig. 10.6 F).

Significance of Meiosis:

(1) During meiotic cell division the chromosome number is reduced to half and due to this the chromosome number in a species remain constant.

(2) Meiosis is playing very important role in reshuffling of genetic material during crossing over. This is useful for the exchange of genetic material (genes) between the homologous chromosomes.

(3) Random separation of chromosomes and crossing over in meiosis are responsible for the variations in plant and animal species.

(4) The genetic variation is the basis raw material for evolution

Comparison between Mitosis and Meiosis:

Mitosis	Meiosis
1. This cell division occurs only once.	1. It involves tow cell divisions.
2. Occurs in somatic cells.	2. Occurs in reproductive cell.
3. It is a short process of 1-2 hrs duration.	3. Meiosis is lengthy process.
4. The genetic material remain constant.	4. The genetic variability is important feature.

5. The chromosome number is kept constant.	5. The chromosome number is reduced to half.
6. Prophase is very simple phase.	6. Prophase I is very complex and having may sequitional phases. (about 4)
7. There is no exchange of chromosomes and crossing over.	7. There is exchange of chromosomes and crossing over.
8. At the end two daughter cells are formed.	8. At the end four daughter cell are formed.

POINTS TO REMEMBER

- The cell cycle is an ordered set of events, culminating in cell growth and division into two daughter cells.
- The cell cycle involves two major phases-growth phase or interphase and mitotic phase (karyokinesis) or Division phase.
- The G_1 phase occupies about 30 – 50% time of the complete cell cycle.
- The S-phase is the synthetic phase of interphase which involves replication of DNA and synthesis of histone proteins which are associated with DNA.
- It occupies about 35 – 45% time of the cell cycle.
- G_2 phase is the second gap or growth phase of the interphase which involves synthesis of ribosomal RNA, messenger RNA and nucleolar RNA.
- Mitosis is the multiplication of the somatic cells into daughter cells of equal size both containing the same number of chromosomes as the parent cell.
- During prophase, the nucleus enlarges and the chromosomes start to condense.
- The chromatids become shorter and thicker; chromosomes are evenly distributed in the nuclear cavity.
- At the end of prophase, the nuclear envelope is rapidly fragmented and disappears releasing nuclear material into the cytoplasm.

- During metaphase, the chromosomes are arranged radially at the equator to form the equatorial plate.
- During anaphase, the chromosomes arranged on equatorial plate divide and the two chromatids of each pair separate.
- During telophase, the chromosomes become less and less condensed and the spindle fibres disappear.
- Cytokinesis is the division of the cytoplasm of a cell following the division of the nucleus.
- In meiosis there are two meiotic divisions meiosis I and meiosis II.
- Meiosis first consists of Prophase I which has about 4 sequential stages.
- Metaphase I, Anaphase I, Telophase I and Cytokinesis.
- Meosis II consists of prophase II, Metaphase II, Anaphase II, Telophase II and Cytokinessis.
- At the end of meiosis four haploid daughter cells are formed.
- Meiosis is the reduction type of cell division in which chromosome number is reduced to half.
- Crossing over and genetic exchange takes place in meiosis.
- Meiosis give rise to variation in species of plants and animals.
- Variation is the basis of evolution.

EXERCISE

1. Give a brief account of cell cycle.
2. Describe somatic cell division or mitosis in an plant cell with suitable diagrams.
3. Draw neat and labelled diagrams to explain mitosis.
4. Give a detailed account of the meiotic cell division and explain its significance.
5. Describe process of meiosis and compare it with mitotic cell division.
6. Describe significance of prophase-I of meiosis.
7. Differentiate between diplotene and pachytene.
8. Describe significance of mitosis and meiosis.
9. Explain the term cell cycle. Explain the different phases in it.
10. Differentiate between mitosis and meiosis.

11. Write short notes:
    a. Significance of meiosis
    b. Cell cycle
    c. Cytokinesis
    d. Zygotene
    e. Pachytene
    f. Diplotene
    g. Significance of mitosis
    h. Anaphase II
    i. Interphase
    j. Metaphase II
    k. G_2 phase

12. Define the terms: cell cycle and mitosis. Name the stages in it.

13. What is Mitosis? Describe the different stages of mitosis.

14. Define mitosis and add a note on its significance.

15. What is meiosis? Add note on its significance.

16. Mention the similarities and differences between mitosis and meiosis.

17. Describe the first meiotic division in brief.

18. Describe the second meiotic division and give its significance.

19. What is meiosis? Describe in detail the process of meiosis.

20. Compare: G_1 and G_2 phase of cell cycle.

21. Explain post mitotic phase.

22. Comment on Early and Late prophase of mitosis.

23. Make list of various stages in meiosis.

24. Explain: (i) Crossing Over, (ii) S-phase, (iii) M-phase, (iv) Chiasmeta, (v) Bivalent, (vi) Tetrads. (vii) Interphase, (viii) Cytokinesis.

11
CHAPTER

INTRODUCTION TO MOLECULAR BIOLOGY

♦ POINTS TO LEARN ♦

11.1 **Introduction**
11.2 **Scope of Molecular biology**
11.3 **Central dogma of Molecular biology**
*   **Points to Remember**
*   **Exercise**

11.1 INTRODUCTION

Molecular biology is a branch of biology that concerns with the molecular basis of biological activities between biomolecules in the various systems of a cell, including the interactions between DNA, RNA, proteins and their biosynthesis, as well as the regulation of these interactions.

Study of molecular biology provides the evidences of the processes of replication, transcription, translation, cell functions and also evolution hence studied worldwide. The central dogma of molecular biology where genetic material is transcribed into RNA and then translated into protein, and proteins perform various functions in the cell.

11.2 SCOPE OF MOLECULAR BIOLOGY

Molecular biology methods have tremendous value not only in the investigations of basic questions, but also in application to a wide variety of problems affecting the overall human kind, animals and plants. Disease prevention and treatment, generation of new protein products, and manipulation of plants and animals for desired phenotypic traits are all applications that are routinely addressed by the application of molecular biology methods. Because of the wide utility of these methods, they are rapidly accepted worldwide.

The study of basic structures and functions of DNA, RNA and proteins give a clue to recognize changes in these biomolecules during any disease or disorder.

DNA replication, DNA damage, repair studies give basic platform to study genetic defects in DNA and defects due to external factors. This study helps to find out remedies in such disorders e.g. Cancer.

Study of transcription, translation and their regulation provides a basic knowledge of the processes and also helps in drug designing to regulate the processes.

Detailed studies on molecular basis of diseases play an important role in finding remedies.

Now-a-days sequencing of genes and proteins has opened new horizons to understand the genetic information and utilization of this information to solve the genetic problems. The sequencing of genomes has also made a new path to study evolutionary relationships at a molecular level.

Bioinformatics is newly emerging discipline where the knowledge of computers and information technology is used in biology. The data generated from molecular biology studies such as DNA-Protein sequences can be stored and processed very easily and available anywhere in the world due to bioinformatics. Some of the *in-vitro* drug studies or studies on model organisms can be replaced by *in-silico* methods (by using computer softwares). This field of molecular biology has played a rewarding role in drug designing and action and also in evolutionary studies.

11.3 CENTRAL DOGMA OF MOLECULAR BIOLOGY

After the discovery of Double helical structure of DNA in 1953, the hypothesis is accepted worldwide that DNA acts as a template for the synthesis of RNA which subsequently determines the arrangement of amino acids within a protein. In 1956, Francis Crick referred this pathway of flow of genetic information as Central Dogma of Molecular biology.

Definition: Central dogma of molecular biology is the flow of genetic information from the DNA to RNA and further to the protein.

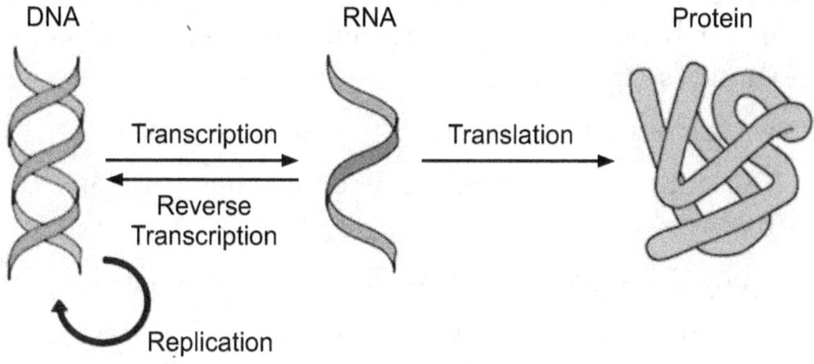

Fig. 11.1: Central Dogma of molecular biology

Here in a given diagram arrows indicate the flow of genetic information. The circular arrow near DNA indicates that the DNA acts as a template for the self-replication of DNA; this is a function of enzyme DNA polymerase. The arrow in between DNA and RNA indicates transcription (copying) of information; during this process DNA acts as a template and RNA polymerase enzyme performs this work. Reverse arrow between DNA and RNA indicates the reverse transcription that means the synthesis of DNA from RNA by the enzyme reverse transcriptase which occurs commonly in case of retroviruses (RNA containing viruses) infection to the host cells. The arrow between RNA and protein suggests the translation of information written in the language of ribonucleotides into the language of amino acids in the protein.

Transcription is the process of making a RNA copy of a gene sequence. This copy, called as messenger RNA (mRNA) molecule, leaves the cell nucleus and enters the cytoplasm, where it directs the synthesis of the protein, which it encodes.

Translation is the process of translating the sequence of a messenger RNA (mRNA) molecule to a sequence of amino acids during protein synthesis. The genetic code describes the relationship between the sequence of base pairs in a gene and the corresponding amino acid sequence that it encodes. In the cell cytoplasm, the ribosome reads the sequence of the mRNA in groups of three bases to assemble the protein.

POINTS TO REMEMBER

- Molecular biology is a branch of biology that concerns with the molecular basis of biological activities between biomolecules in the various systems of a cell.

- Molecular biology methods have tremendous scope in:
  - The study of basic structures and functions of DNA, RNA and proteins.
  - Study of transcription, translation and their regulation.
  - To study genetic defects in DNA, to find remedies on diseases on molecular basis.
  - Sequencing of genes and proteins and Bioinformatics.
  - Central dogma of molecular biology is the flow of genetic information from the DNA to RNA and further to the protein.

- **Transcription** is the process of making a RNA copy of a gene sequence.

- **Translation** is the process of translating the sequence of a messenger RNA (mRNA) molecule to a sequence of amino acids during protein synthesis.

EXERCISE

1. What is Molecular biology?
2. Give central dogma of molecular biology?
3. What is Transcription?
4. What is Translation?
5. Explain the scope of Molecular biology?
6. Answer in one sentence
   (a) What is the function of DNA polymerase enzyme?
   (b) RNA polymerase enzyme
   (c) Reverse transcriptase enzyme

12

CHAPTER

STRUCTURE OF DNA

♦ POINTS TO LEARN ♦

12.1 Structure of DNA

12.2 Nucleotides and Nucleosides

12.3 Chargaff's Rule

12.4 C-value paradox

***   Points to Remember**

***   Exercise**

12.1 STRUCTURE OF DNA

Deoxyribonucleic acid (DNA) acts as a genetic material in most of the living organisms except some RNA (retro) viruses. Fragment of DNA is called as gene which is considered as a single unit of heredity. The sequence of a gene transcribed into mRNA and then translated into protein. Damage in the DNA can be repaired using DNA repair system. DNA is a polymer of four different nucleotides (dATP, dGTP, dCTP, dTTP). Nucleotides also have some other functions like as energy carriers: ATP, GTP; signal transduction: cyclic AMP; coenzymes: NAD, FAD etc.

12.2 NUCLEOTIDES AND NUCLEOSIDES

A nucleoside is made up of a pentose sugar and nitrogenous base while a nucleotide is made up of a pentose sugar, nitrogenous base and a phosphate group. DNA consists of deoxyribonucleotides while RNA consists of ribonucleotides.

1. Sugar: Deoxyribose is a cyclic pentose (5-carbon) sugar occurs in DNA while sugar in RNA is a ribose. Carbons in the sugar are noted as 1' to 5'. A nitrogen atom of the nitrogenous base binds to C1' of a sugar by N-glycosidic bond and the phosphate group binds to C5' by phospho-ester bond to make the nucleotide. The nucleotide is therefore, phosphate - C5' sugar C1' - nitrogen base.

(12.1)

Deoxyribose in DNA
(pentose: sugar with 5 carbons)

Ribose in RNA
(another pentose)

Orientation of the DNA (5' → 3'):
according to the carbon of the pentose

Fig. 12.1: Chemical Structure of Sugars in DNA and RNA

2. Nitrogenous bases: Nitrogen bases are nitrogen containing aromatic heterocyclic compounds. Purines and pyrimidines are two different types of nitrogen bases found in DNA and RNA. Adenine (A) and Guanine (G) are purines while Cytosine (C), Thymine (T) and Uracil (U) are pyrimidines. (Note: thymine is replaced by uracil in RNA).

Nitrogen bases: General structure

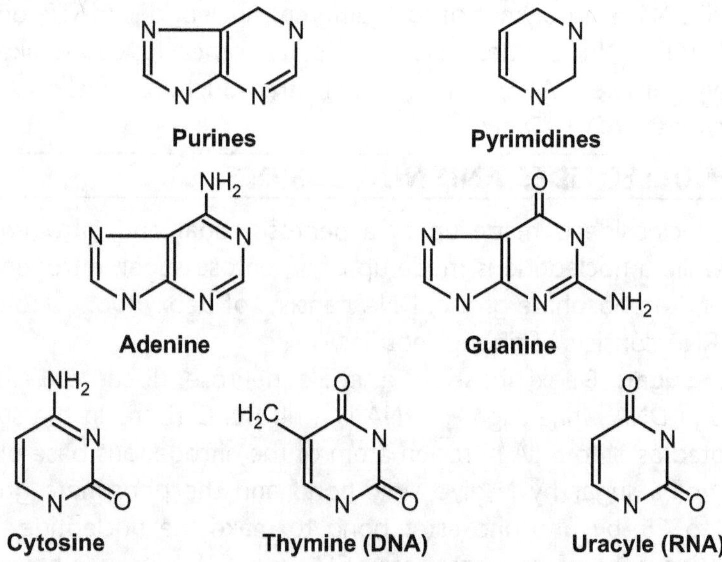

Purines

Pyrimidines

Adenine

Guanine

Cytosine

Thymine (DNA)

Uracyle (RNA)

Fig. 12.2: Chemical Structure

Table 12.1

RNA: Nucleosides and nucleotides:

S. N.	Name of Nitrogen base	Name of a pentose Sugar	Name of Nucleoside	Number of Phosphate groups	Name of Nucleotide
1.	Adenine	Ribose	Adenosine	One	Adenosine Mono Phosphate (AMP) (Adenylate)
				Two	Adenosine Diphosphate (ADP)
				Three	Adenosine Triphosphate (ATP)
2.	Guanine	Ribose	Guanosine	One	Guanosine Mono Phosphate (GMP) (Guanylate)
				Two	Guanosine Diphosphate (GDP)
				Three	Guanosine Triphosphate (GTP)
3.	Cytosine	Ribose	Cytidine	One	Cytidine Mono Phosphate (CMP) (Cytidylate)
				Two	Cytidine Diphosphate (CDP)
				Three	Cytidine Triphosphate (CTP)
4.	Uracil	Ribose	Uridyl	One	Uridyl Mono Phosphate (UMP) (Uridylate)
				Two	Uridyl Diphosphate (UDP)
				Three	Uridyl Triphosphate (UTP)

Table 12.2

DNA: Nucleosides and nucleotides:

S.N.	Name of Nitrogen base	Name of a pentose Sugar	Name of Nucleoside	No. of Phosphate groups	Name of Nucleotide
1.	Adenine	Deoxyribose	Adenosine	One	Deoxyadenosine Mono Phosphate (dAMP) (deoxyadenylate)
				Two	Deoxyadenosine Diphosphate (dADP)
				Three	Deoxyadenosine Triphosphate (dATP)

Contd...

2.	Guanine	Deoxyribose	Guanosine	One	Deoxyguanosine Mono Phosphate (dGMP) (deoxyguanylate)
				Two	Deoxyguanosine Diphosphate (dGDP)
				Three	Deoxyguanosine Triphosphate (dGTP)
3.	Cytosine	Deoxyribose	Cytidine	One	Deoxycytidine Mono Phosphate (dCMP) (deoxycytidylate)
				Two	Deoxycytidine Diphosphate (dCDP)
				Three	Deoxycytidine Triphosphate (dCTP)
4.	Thymine	Deoxyribose	Thymidine	One	Deoxythymidine Mono Phosphate (dTMP) (deoxythymidylate)
				Two	Deoxythymidine Diphosphate (dTDP)
				Three	Deoxythymidine Triphosphate (dTTP)

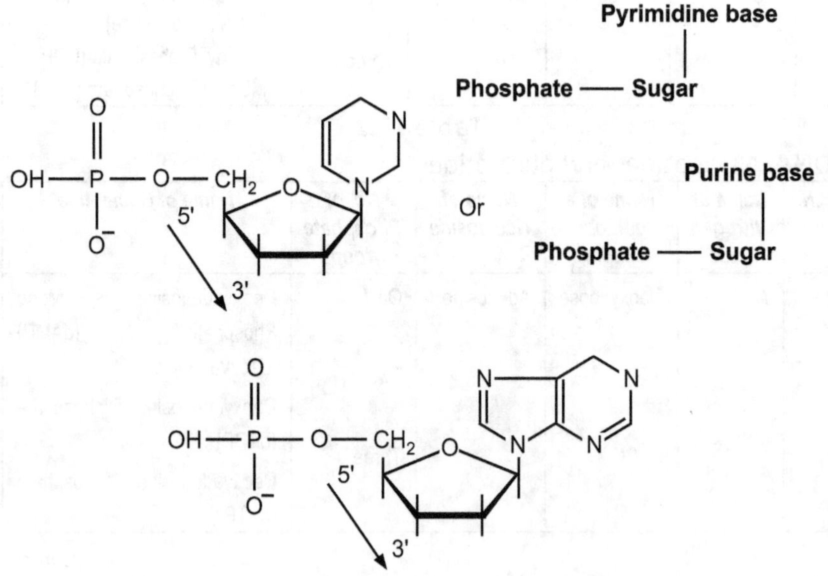

Pyrimidine base

Phosphate —— Sugar

Purine base

Or

Phosphate —— Sugar

3. **Phosphoric acid:** It gives a phosphate group.

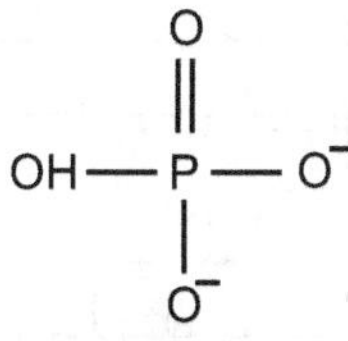

Phosphate

12.3 CHARGAFF'S RULE

Chargaff's rule states that DNA from any cell of any organism has a 1 : 1 ratio of pyrimidine and purine bases and, more specifically, that the amount of guanine, a purine base, is equal to cytosine, a pyrimidine base; and the amount of adenine, a purine base, is equal to thymine, a pyrimidine base.

So a base pair is composed of a pyrimidine base and a purine base. This pattern is found in both strands of DNA, and is responsible for the base-pairing rule, which states that adenine always pairs with thymine, and guanine always pairs with cytosine.

1. The total amount of adenine released is equal to the total amount of thymine and similarly, the total amount of cytosine is equal to the total amount of guanine.

   i.e. $A = T$ and $C = G$

2. It also states that in natural DNA the base ration A/T is closed to unity (= 1) and C/G is also close to unity (= 1)

   i.e. $(A/T) = (C/G) = 1$

3. $A + G = T + C$ and

4. $A + C = G + T$

12.4 C VALUE PARADOX

The C-value is the amount of DNA in the haploid genome of an organism. It varies over a very wide range, with a general increase in C-value with complexity of organism from prokaryotes to invertebrates, vertebrates, plants. C-value paradox is the paradox in amount of DNA and complexity of organism. Eukaryotic genome size fails to correlate with complexity of organism. The largest genome is found in an Amoeba, a one-cell organism, with 686,000 Mb, 200 fold larger than the human genome and 20,000 fold larger than the one found in Yeast.

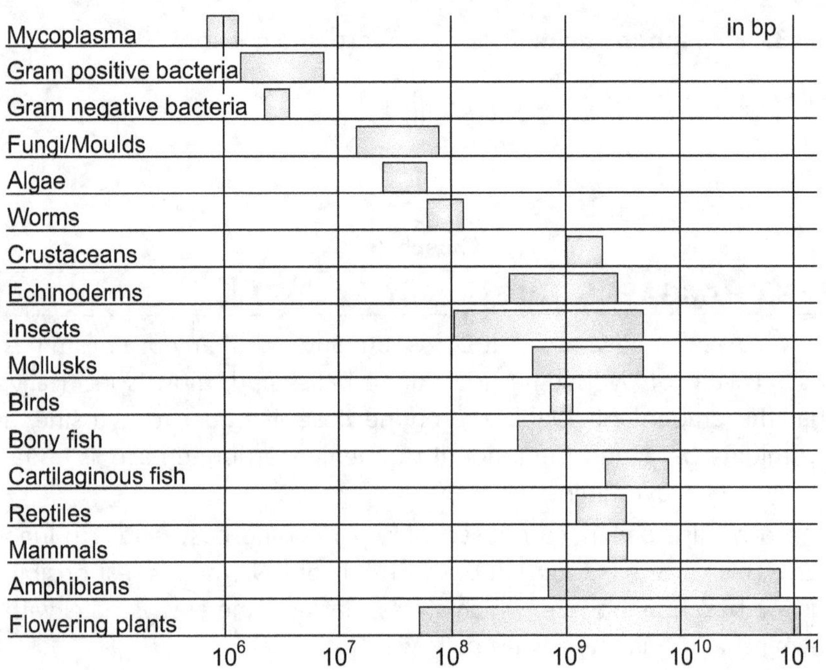

Fig. 12.3: C-value paradox in different groups of plants and animals

Our current understanding of complex genomes reveals several factors that help to explain the classic C-value paradox:

o   Introns in genes

o   Regulatory elements of genes

o   Pseudogenes

o   Multiple copies of genes

o   Intergenic sequences

o   Repetitive DNA

The facts that some of the genomic DNA from complex organisms is highly repetitive, and that some proteins are encoded by families of genes whereas others are encoded by single genes, mean that the genome can be considered to have several distinctive components. Analysis of the kinetics of DNA reassociation, largely in the 1970's, showed that such genomes have components that can be distinguished by their repetition frequency.

POINTS TO REMEMBER

- Deoxyribonucleic acid (DNA) acts as a genetic material in most of the living organisms.
- A nucleoside is made of a pentose sugar and nitrogenous base.
- A nucleotide is made of a pentose sugar, nitrogenous base and a phosphate group.
- Deoxyribose is a cyclic pentose (5-carbon) sugar in DNA while sugar in RNA is a ribose.
- Nitrogen bases are nitrogen containing aromatic heterocyclic compounds.
- Purines and pyrimidines are two different types of nitrogen bases found in DNA and RNA.
- Adenine (A) and Guanine (G) are purines while Cytosine (C), Thymine (T) and Uracil (U) are pyrimidines.
- Chargaff's rule states that DNA from any cell of any organism has a 1 : 1 ratio of pyrimidine and purine bases.
- The C-value is the amount of DNA in the haploid genome of an organism. C-value paradox is the paradox in amount of DNA and complexity of organism.

EXERCISE

1. Describe the Structure of DNA.
2. Explain Structure of nitrogen bases.
3. What is nucleoside?
4. What is nucleotide?
5. Discuss Chargaff's rule.
6. Describe C-value paradox.
7. Name Purine bases.
8. Name Pyrimidine bases.
9. What is pentose sugar?
10. What is deoxyribose sugar?
11. Explain the pairing of nitrogen bases in DNA and RNA.
12. Which base is replaced in RNA and by what?

13
CHAPTER

WATSON CRICK MODEL OF DNA

◆ POINTS TO LEARN ◆

13.1 Introduction
13.2 Characteristics of DNA
13.3 Types of DNA (A, B and Z DNA).
    13.3.1 B-form DNA
    13.3.2 A-form of DNA
    13.3.3 Z-form DNA
*   **Points to Remember**
*   **Exercise**

13.1 INTRODUCTION

DNA is the largest macromolecule that represents the genetic material of the cell. Chemically, DNA is a double helix of two antiparallel polynucleotide chains. Each polynucleotide chain is a linear mixed polymer of four deoxyribotides i.e. deoxyadenylate, deoxyguanylate, deoxycytidylate and deoxythymidylate.

In 1953, American Biologist J.D. Watson and British Physicist F.H.C. Crick proposed the three-dimensional model of physiological DNA (i.e B-DNA) on the basis of X-ray diffraction data of DNA obtained by Franklin and Wilkins. For this epoch-making discovery, Watson, Crick and Wilkins got Nobel Prize in medicine in 1962. Term DNA was given by Zaccharis.

The important features of Watson – Crick Model or double helix model of DNA are as follows:

1. The DNA molecule consists of two polynucleotide chains or strands that spirally twisted around each other and coiled around a common axis to form a right-handed double-helix.

2. The two strands are antiparallel i.e. they run in opposite directions so that the 3' end of one chain facing the 5' end of the other.

3. The sugar-phosphate backbones remain on the outside, while the core of the helix contains the purine and pyrimidine bases.

4. The two strands are held together by hydrogen bonds between the purine and pyrimidine bases of the opposite strands.

5. The base compositions of DNA obey Chargaff's rule (E.E. Chargaff, 1950) according to which A pairs with T by 2 hydrogen bonds and G pairs with C by 3 hydrogen bonds; that means $\sum$ purines (A+G) = $\sum$ pyrimidines (C+T); also (A+C) = (G+T). It also states that ratio of (A+T) and (G+C) is constant for a species (range 0.4 to 1.9).

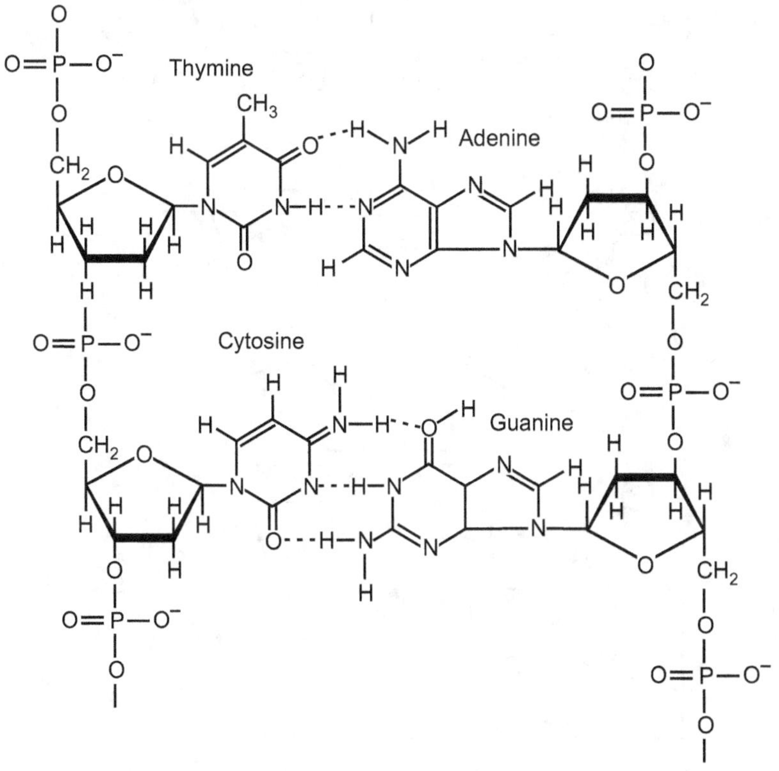

Fig. 13.1: Base composition of DNA obey Chargaff's rule

[A : pairs with T and G pairs with G)

6. The diameter of DNA is 20 nm (20 Å). Adjacent bases are separated by 0.34 nm (3.4 Å) along the axis. The length of a complete turn of helix is 3.4 nm (34 Å) i.e. there are 10 bps per turn.

7. The DNA helix has a shallow groove called minor groove and a deep groove called major groove across the length.

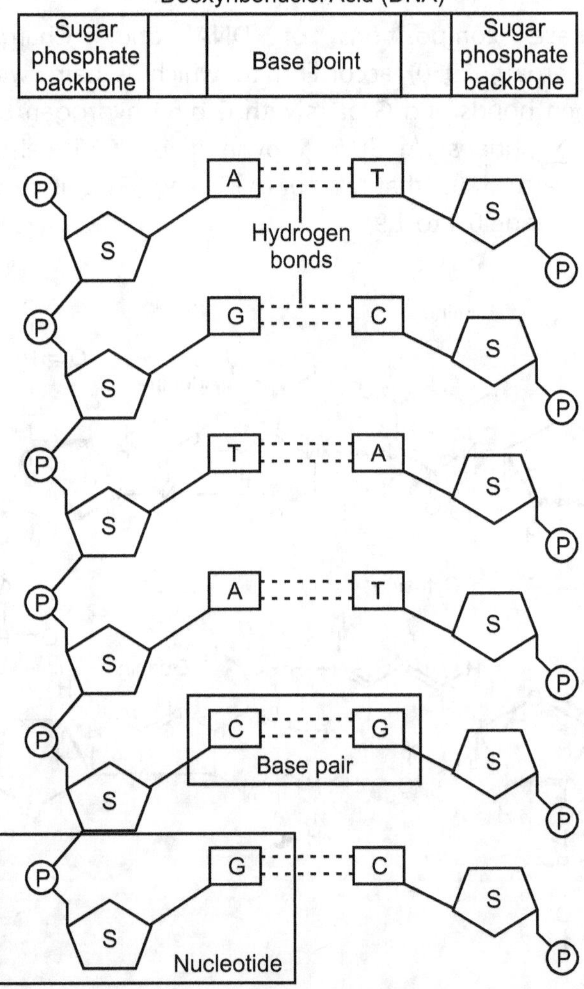

Fig. 13.2: Structure of DNA showing sugar phosphate backbone with base paring by hydrogen bonds

Nitrogenous Bases

G Guanine C Cytosine

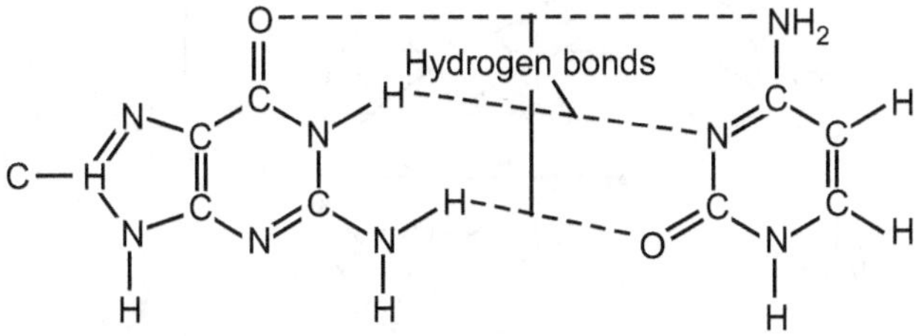

A Adenine T Thymine

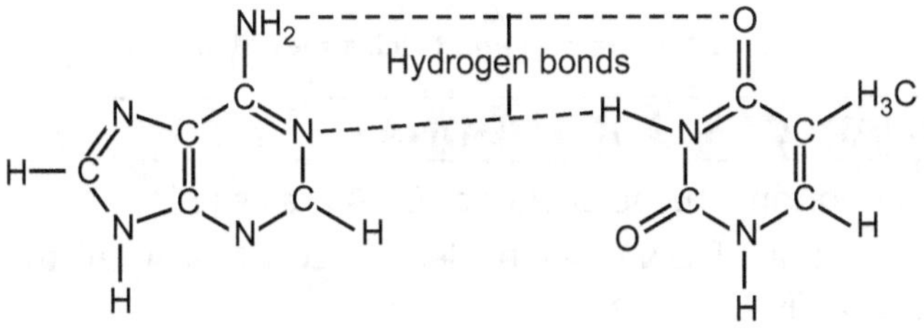

U Uracil

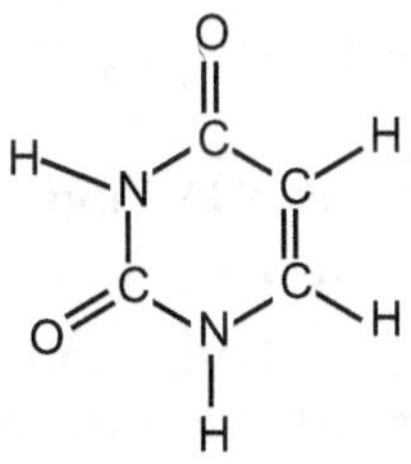

Replaces Thymine in RNA

Fig. 13.3: Molecular structure of DNA and pairing of nitrogen bases in DNA

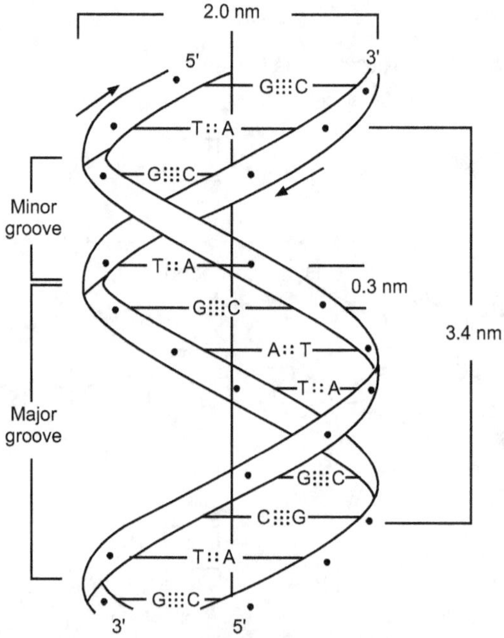

Fig. 13.4: Watson and Crick's model of DNA

13.2 CHARACTERISTICS OF DNA

The important characteristics of DNA are as follows:

1. The amount of DNA per nucleus is constant in all the somatic cells of a given species.

2. The total amount of DNA in a haploid genome is a characteristic of each organism and is known as C-value.

3. Only a small fraction of DNA is functional in eukaryotes.

4. DNA is the chemical basis of heredity and is organized into genes or cistrons.

5. DNA replicates to form DNAs and transcribes to form RNAs.

6. DNA replication occurs in the S-phase of cell cycle.

7. DNA replication is semi-conservative in which two daughter DNA molecules formed; each receives one of parental strand and one new strand.

8. One strand of DNA directs the synthesis called template strand, or antisense or non-coding strand. The other strand is called coding or non-template or sense strand which has the same sequence as the RNA transcript except T in place of U.

9. DNA has many repeated base sequences, commonly called as satellite DNA.

10. DNA can easily undergo denaturation (melting) and renaturation with any change in pH, temperature and salt concentration. DNA with a high G + C content is more resistant to thermal melting than A + T rich molecules.

11. DNA can be synthesized *in-vitro* (in the laboratory).

12. DNA can be measured by the unit picogram (1pg = 10 g)

3.3 TYPES OF DNA (A, B AND Z DNA).

After the discovery of DNA double helix in 1953, DNA was thought to have the typical B form (Watson and Crick model) everywhere in all conditions. But later on after around 20 years, various experiments led the discovery of different forms of DNA. Thus, DNA can be of the following types; A-form, B-form and Z-form.

3.3.1 B-form DNA

It is biologically important form of DNA which is naturally found in most living organisms. This form of DNA can be found in other different forms such as A, Z-forms in conditions like relative humidity, ionic strength, salt concentration and sequence of nucleotides. Other forms differ from B-form in features such as number of bp per turn, distance between two bp, diameter etc.

Note: Features of B form are discussed earlier in Watson and Crick model of DNA.

3.3.2 A-form of DNA

A-form of DNA is right handed but less hydrated than B-form of DNA. It has 11bp per turn with a shorter distance between adjacent bases. It is bigger in diameter than B-DNA i.e. 23 Å. The bases are tilted more in relation to the axis of the helix than the B-DNA. A-DNA may occur under experimental conditions, also RNA-DNA duplexes and RNA-RNA duplexes show A-form.

3.3.3 Z-form DNA

Z-DNA is a radically different duplex structure, with the two strands coiling in left-handed helices and a pronounced zig-zag (hence the name) pattern in the phosphodiester backbone. Z-DNA can form when the DNA is in an alternating purine-pyrimidine sequence such as GCGCGC, and indeed the G and C nucleotides are in different conformations, leading to the zig-zag pattern. Z-DNA can also found in solutions of high ionic strength, such as 2M NaCl. The helix of Z-DNA is 18 Å in diameter, containing 12 base pairs per turn.

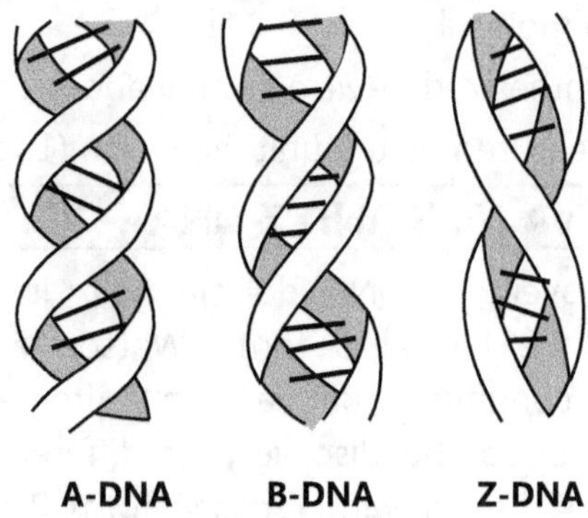

Fig. 13.5: Different types of DNA

Even classic B-DNA is not completely uniform in its structure. X-ray diffraction analysis of crystals of duplex oligonucleotides shows that a given sequence will adopt a distinctive structure. These variations in B-DNA may differ in the propeller twist (between bases within a pair) to optimize base stacking, or in the three ways that two successive base pairs can move relative to each other: twist, roll, or slide.

Characteristics	B-Form	A-Form	Z-Form
Conditions	92% relative humidity, low ionic strength	75% relative humidity, presence of Na^+, K^+, Cs^+ ions	Very high ionic strength
Helix sense	Right Handed	Right Handed	Left Handed
Base pairs per turn	10	11	12
Vertical rise per bp	3.4 Å	2.56 Å	3.71 Å
Rotation per bp	+36°	+33°	-30°
Helix diameter	20 Å	23 Å	18 Å

POINTS TO REMEMBER

- DNA is the largest macromolecule that represents the genetic material of the cell.
- The important features of Watson – Crick Model.
- The DNA molecule consists of two polynucleotide chains, which are antiparallel.
- The two strands are held together by hydrogen bonds between the purine and pyrimidine bases.
- The base compositions of DNA obey Chargaff s rule. The DNA helix has a shallow groove called minor groove and a deep groove called major groove across the length.
- The DNA is of three types; A-form, B-form and Z-form.

Characteristics	B-Form	A-Form	Z-Form
Helix sense	Right Handed	Right Handed	Left Handed
Base pairs per turn	10	11	12
Vertical rise per bp	3.4 Å	2.56 Å	3.71 Å
Rotation per bp	+36°	+33°	-30°
Helix diameter	20 Å	23 Å	18 Å

EXERCISE

1. Give an account of Watson Crick model of DNA.
2. Describe characteristic features of Watson Crick model of DNA.
3. Explain different types of DNA.
4. Write short notes on:
   - (a) Types of DNA
   - (b) A DNA
   - (c) B DNA
   - (d) Z DNA
5. Distinguish between:
   - (a) A DNA and Z DNA
   - (b) A DNA and B DNA
6. Sketch and label the structure of DNA.

14

CHAPTER

PACKING OF DNA INTO CHROMOSOMES

♦ POINTS TO LEARN ♦

14.1   **Introduction**
14.2   **Packaging of eukaryotic DNA into a set of Chromosomes**
14.3   **Types of Chromosomes**
*   **Points to Remember**
*   **Exercise**

14.1 INTRODUCTION

The most important function of DNA is to carry genes, the information that specifies all the proteins that make up an organism including information about when, in what types of cells, and in what quantity each protein is to be made. The genomes of eukaryotes are divided into chromosomes, and in this section we see how genes are typically arranged on each chromosome. In addition, we describe the specialized DNA sequences that allow a chromosome to be accurately duplicated and passed on from one generation to the next.

We also confront the serious challenge of DNA packaging. Each human cell contains approximately 2 meters of DNA if stretched end-to-end; yet the nucleus of a human cell, which contains the DNA, is only about 6 μm in diameter. This is geometrically equivalent to packing 40 km (24 miles) of extremely fine thread into a tennis ball. The complex task of packaging DNA is accomplished by specialized proteins that bind to and fold the DNA, generating a series of coils and loops that provide increasingly higher levels of organization, preventing the DNA from becoming an unmanageable tangle. Amazingly, although the DNA is very tightly folded, it is compacted in a way that allows it to easily become available to many enzymes in

(14.1)

the cell that replicate it, repair it, and use its genes to produce proteins.

Prokaryotic packaging of DNA:

Bacteria carry their genes on a single DNA molecule, which is usually double stranded covalently closed and circular. This DNA is associated with proteins those perform packaging and condensation of the DNA, but they are different from the proteins that perform these functions in eukaryotes. Although often called the bacterial "chromosome," it does not have the same structure as eukaryotic chromosomes.

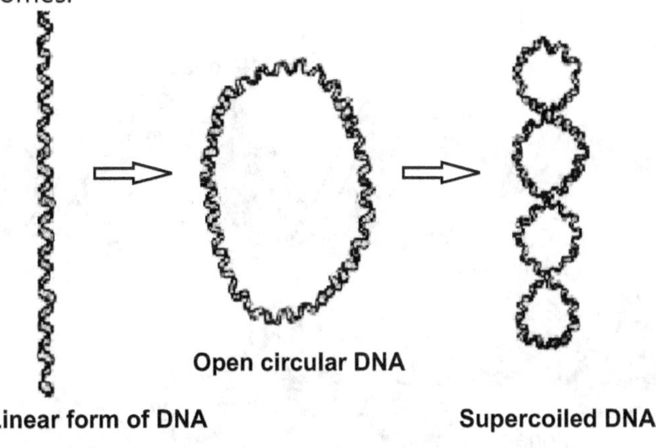

Fig. 14.1: Condensation of bacterial DNA

14.2 PACKAGING OF EUKARYOTIC DNA INTO A SET OF CHROMOSOMES

Every human cell contains over two metres of DNA, packaged into a nucleus of 5-20 μm in diameter. To achieve this, the DNA is packaged in a highly ordered process. This allows condensation of the DNA but still enables replication and transcription machinery access. Several groups of proteins are involved in the condensation of DNA.

Levels of DNA condensation within a eukaryotic chromosome:

1. The first level of DNA condensation occurs when 146 bp of DNA is wrapped around histone proteins named as nucleosome. DNA is wrapped twice around a histone core to form the beads-on-a-string level of packing termed as

chromatin. This produces a fibre approximately 11 nm in width.

2. The second level of chromatin condensation is the arrangement of the nucleosomes into a 30 nm chromatin fibre where six nucleosomes per turn are found; this structure is termed as solenoid.

3. The chromatin fibre is then arranged in loop domains each approximately 120 Kb long. These chromatin loops form rosettes. Each rosette is termed as multi-loop sub-compartment (MLS) and comprises of six loops.

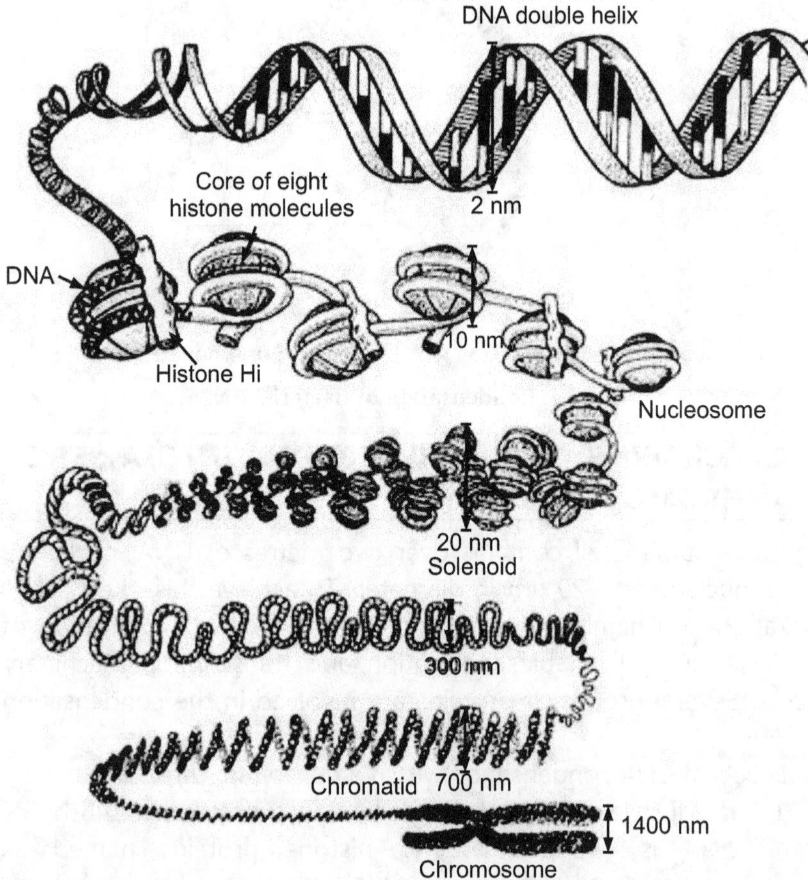

Fig. 14.2: Solenoid model of folding and super folding of basic chromatin components to form eukaryotic chromosome

4. A further 120 Kb of DNA links each rosette. The rosette is attached at its centre to the nuclear matrix and is associated with Histone H1.

5. The final level of chromosome compaction occurs prior to cell division in mitosis. Mitotic chromosomes have a diameter of 700 nm. The rosettes are lined up and attached to a central chromosome scaffold of non-histone acidic proteins

14.3 TYPES OF CHROMOSOMES

A) On the basis of number of centromeres:
1. Monocentric with one centromere.
2. Dicentric with two centromeres, one in each chromatid.
3. Polycentric with more than two centromeres.
4. Acentric without centromere. Such chromosomes represent freshly broken segments of chromosomes, which do not survive for long.
5. Diffused or non-located centromeres are distributed throughout the length of chromosome. The microtubules of spindle fibres are attached to chromosome arms at many points. The diffused centromeres are found in insects, some algae and some groups of plants.

B) Based on the position of centromere:
1. **Telocentric:** These are rod-shaped chromosomes with centromere occupying a terminal position. One arm is very long and the other is absent.
2. **Acrocentric:** These are rod-shaped chromosomes having sub-terminal centromere. One arm is very long and the other is very small.
3. **Sub metacentric:** These are J or L-shaped chromosomes with centromere slightly away from the mid-point so that the two arms are unequal.
4. **Metacentric:** These are V-shaped chromosomes in which centromere lies in the middle of chromosomes so that the two arms are almost equal.

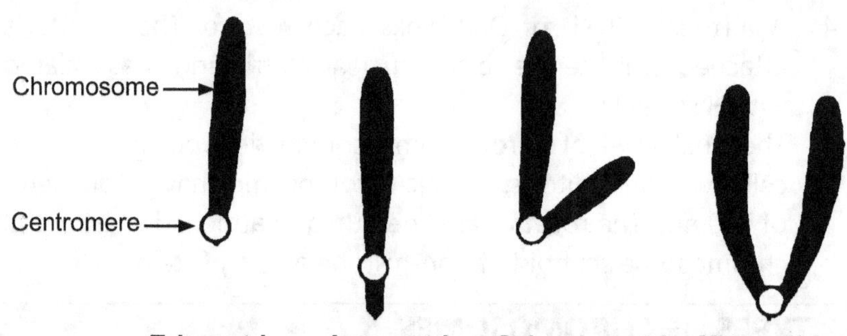

Fig. 4.4: Types of Chromosomes

POINTS TO REMEMBER

- The most important function of DNA is to carry genes.

- The genomes of eukaryotes are divided into chromosomes.

- The complex task of packaging DNA is accomplished by specialized proteins.

- DNA molecule which is associated with packaging and condensation.

- On the basis of number of centromeres chromosomes types are Monocentric, Dicentric, Polycentric, Acentric.

- Based on the position of centromere chromosomes types are Telocentric, Acrocentric, Sub-metacentric, and Metacentric.

EXERCISE

1. Explain packaging of Eukaryotic DNA into chromosomes.

2. Describe different types of chromosomes based on the position of centromere.

3. Write short notes on:

    (a) Types of chromosomes based on the position of centromere.

    (b) Types of chromosomes based on the number of centromere.

    (c) Packaging of Eukaryotic DNA into chromosomes.

4. Explain:

   (i)　Polycentric and dicentric chromosomes.

   (ii)　Telocentric and meta centric chromosomes.

   (iii)　Functions of DNA.

   (iv)　Packaging of DNA in Prokaryotic cell.

   (v)　Packaging of DNA in Eukaryotic cell.

15

CHAPTER

STRUCTURE AND TYPES OF RNA

◆ POINTS TO LEARN ◆

15.1 **Introduction**
15.2 **RNA Structure**
15.3 **Functions of RNA in Protein Synthesis**
15.4 **RNA as Hereditary Information**
* **Points to Remember**
* **Exercise**

15.1 INTRODUCTION

Expression of the information in a gene generally involves production of an RNA molecule transcribed from a DNA template. Strands of RNA and DNA may seen quite similar at first glance, differing only in that RNA has a hydroxyl group at the 2′ position of the ribose sugar and uracil instead of thymine. However, unlike DNA, most RNAs carry out their functions as single strands, strands that fold back on themselves and have the potential for much greater structural diversity than DNA. RNA is thus suited to a variety of cellular functions.

15.2 RNA STRUCTURE

RNA is typically single stranded and is made of ribonucleotides that are linked by phosphodiester bonds. A ribonucleotide in the RNA chain contains ribose (the pentose sugar), one of the four nitrogenous bases (A, U, G, and C), and a phosphate group. The subtle structural difference between the sugars gives DNA added stability, making DNA more suitable for storage of genetic information, whereas the relative instability of RNA makes it more suitable for its more short-term functions. The RNA-specific pyrimidine uracil forms a complementary base pair with adenine and

(15.1)

is used instead of the thymine used in DNA. Even though RNA is single stranded, most types of RNA molecules show extensive intra-molecular base pairing between complementary sequences within the RNA strand, creating a predictable three-dimensional structure essential for their function (Fig. 15.1 and Fig. 15.2).

Deoxyribose (in DNA)   Ribose (in RNA)

(a) Ribonucleotides contain the pentose sugar ribose instead of the deoxyribose found in deoxyribonucleotides

Thymine (in DNA)
T

Uracil (in RNA)
U

(b) RNA contains the pyrimidine uracil in place of thymine found in DNA

Fig. 15.1: Sugars and nitrogen bases in RNA

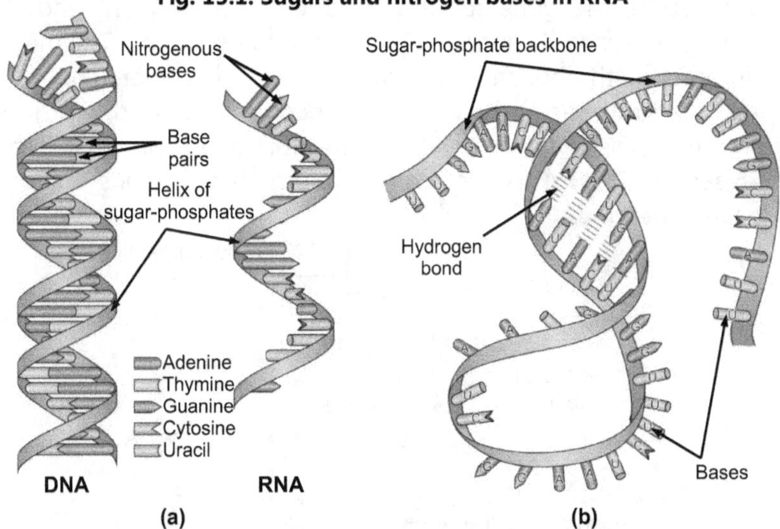

Fig. 15.2: (a) DNA is typically double stranded, whereas RNA is typically single stranded. (b) Although it is single stranded, RNA can fold upon itself, with the folds stabilized by short areas of complementary base pairing within the molecule, forming a three-dimensional structure

15.3 FUNCTIONS OF RNA IN PROTEIN SYNTHESIS

Cells access the information stored in DNA by creating RNA to direct the synthesis of proteins through the process of translation. Proteins within a cell have many functions, including building cellular structures and serving as enzyme catalysts for cellular chemical reactions that give cells their specific characteristics. The three main types of RNA directly involved in protein synthesis are messenger RNA (mRNA), ribosomal RNA (rRNA), and transfer RNA (tRNA).

In 1961, French scientists François Jacob and Jacques Monod hypothesized the existence of an intermediary between DNA and its protein products, which they called messenger RNA. Evidence supporting their hypothesis was gathered soon afterwards showing that information from DNA is transmitted to the ribosome for protein synthesis using mRNA. If DNA serves as the complete library of cellular information, mRNA serves as a photocopy of specific information needed at a particular point in time that serves as the instructions to make a protein.

mRNA structure and functions:

The mRNA carries the message from the DNA, which controls all of the cellular activities in a cell. If a cell requires a certain protein to be synthesized, the gene for this product is "turned on" and the mRNA is synthesized through the process of transcription. The mRNA then interacts with ribosomes and other cellular machinery (Fig. 15.3) to direct the synthesis of the protein it encodes during the process of translation. The mRNA is relatively unstable and short-lived in the cell, especially in prokaryotic cells, ensuring that proteins are only made when needed.

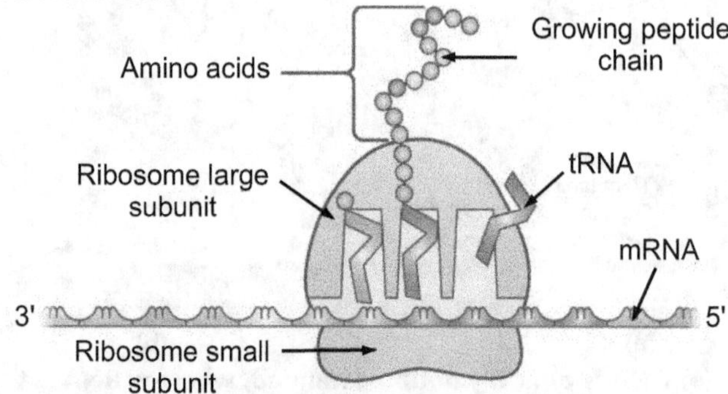

Fig. 15.3: A generalized illustration of how mRNA and tRNA are used in protein synthesis within a cell

The rRNA and tRNA are stable types of RNA. In eukaryotes, synthesis, cutting, and assembly of rRNA into ribosomes takes place in the nucleolus region of the nucleus, but these activities occur in the cytoplasm of prokaryotes. Neither of these types of RNA carries instructions to direct the synthesis of a protein, but they play other important roles in protein synthesis.

rRNA structure and functions:

Ribosomes are composed of rRNA and ribonucleoproteins (RNPs). As its name suggests, rRNA is a major constituent of ribosomes, composing up to about 60% of the ribosome by mass and providing the location where the mRNA binds. The rRNA ensures the proper alignment of the mRNA, tRNA, and the ribosomes; the rRNA of the ribosome also has an enzymatic activity (peptidyl transferase) and catalyzes the formation of the peptide bonds between two aligned amino acids during protein synthesis. Although rRNA had long been thought to serve primarily a structural role, its catalytic role within the ribosome was proven in 2000.

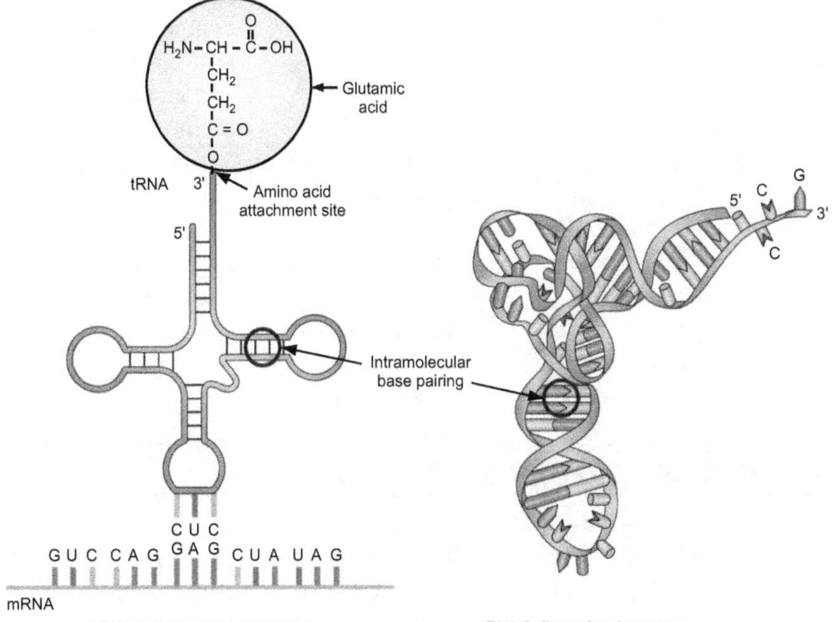

Fig. 15.4: A tRNA molecule is a single-stranded molecule that exhibits significant intracellular base pairing, giving it its characteristic three-dimensional shape

tRNA Structure and Functions:

Transfer RNA is the third main type of RNA and one of the smallest, usually only 70–90 nucleotides long. It carries the correct amino acid to the site of protein synthesis in the ribosome. It is the base pairing between the tRNA and mRNA that allows for the correct amino acid to be inserted in the polypeptide chain being synthesized (Fig. 15.4). Any mutations in the tRNA or rRNA can result in global problems for the cell because both are necessary for proper protein synthesis (Table 15.1).

Table 15.1: Structure and Functions of RNA

	mRNA	**rRNA**	**tRNA**
Structure	Short, unstable, single-stranded RNA corresponding to a gene encoded within DNA	Longer, stable RNA molecules composing 60% of ribosome's mass	Short (70-90 nucleotides), stable RNA with extensive intramolecular base pairing; contains an amino acid binding site and an mRNA binding site
Function	Serves as intermediary between DNA a nd protein; used by ribosome to direct synthesis of protein it encodes	Ensures the proper alignment of mRNA, tRNA and ribosome during protein synthesis; catalyzes peptide bond formation between amino acids	Carries the correct amino acid to the site of protein synthesis in the ribosome

15.4 RNA AS HEREDITARY INFORMATION

Although RNA does not serve as the hereditary information in most cells, RNA does hold this function for many viruses that do not contain DNA. Thus, RNA clearly does have the additional capacity to serve as genetic information. Although RNA is typically single stranded within cells, there is significant diversity in viruses. Human Immuno-deficiency Virus (HIV) which causes AIDS, Rhinoviruses, which cause the common cold; influenza viruses; and the Ebola virus, are single-stranded RNA virus. Rotaviruses, which cause severe gastroenteritis in children and other immune compromised individuals, are examples of double-stranded RNA viruses. Because double-stranded RNA is uncommon in eukaryotic cells, its presence serves as an indicator of viral infection.

POINTS TO REMEMBER

- RNA is typically single stranded and is made of ribonucleotides that are linked by phosphodiester bonds.
- RNA contains four nitrogenous bases i.e. A, U, G, and C.
- The information stored in DNA is transcribed to RNA which is directed the synthesis of proteins.
- Proteins build cellular structures and serve as an enzyme catalysts in cells.
- The mRNA rRNA and tRNA are directly involved in protein synthesis.
- DNA serves as the complete library of cellular information.
- mRNA serves as a photocopy of specific information.
- The mRNA carries the message from the DNA, which controls all of the cellular activities in a cell.
- Ribosomes are composed of rRNA and ribonucleo proteins (RNPs).
- The rRNA ensures the proper alignment of the mRNA, tRNA, and the ribosomes.
- The tRNA carries the correct amino acid to the site of protein synthesis in the ribosome.

EXERCISE

1. Describe Structure of RNA.
2. Explain different types of RNAs.
3. Write short notes on:
   - (a) Types of RNA
   - (b) mRNA
   - (c) tRNA
   - (d) rDNA
4. Compare and give differences between mRNA and tRNA.
5. Compare and give difference between rRNA and mRNA.
6. Comment on: RNA as hereditary material.
7. Sketch and label:
   - (a) Structure of RNA
   - (b) Structure of mRNA
   - (c) Structure of rRNA
   - (d) Structure of tRNA

16

CHAPTER

DNA REPLICATION

♦ POINTS TO LEARN ♦

16.1   DNA Replication
16.2   Types of DNA Replication
*   Point to Remember
*   Exercise

16.1 DNA REPLICATION

Long before the structure of DNA was known, scientists wondered at the ability of organisms to create faithful copies of themselves and, later, at the ability of cells to produce many identical copies of large and complex macromolecules. There should be a template, a structure that would allow molecules to be lined up in a specific order and joined, to create a macromolecule with a unique sequence and function. The 1940s brought the revelation that DNA was the genetic molecule, but not until James Watson and Francis Crick deduced its structure did the way in which DNA could act as a template for the replication and transmission of genetic information become clear: one strand is the complement of the other. So the replication is copying of sequence of parent DNA strand into daughter strand.

16.2 TYPES OF DNA REPLICATION

Watson and Crick had proposed that in order to copy itself, DNA would have to open down the center, sort of like a zipper coming apart, so that a new DNA strand could be built on top of the exposed strands. Following the rules of complimentary base pairing, adenine would pair with thymine, and cytosine would pair with guanine. This idea was called a template model, since one DNA strand serves as the template for a new one.

(16.1)

Watson and Crick figured that this model would result in two new double strands of DNA, each one with one strand of parent (or template) DNA and one strand of daughter (or newly-synthesized) DNA. They called this the semi-conservative model, because half of the parent DNA was conserved in each new DNA molecule.

Scientists looked at the double helix of DNA and wondered how in the world it could possibly open itself up without getting tangled or torn apart. So they thought up some other ideas about how DNA replication works. One hypothesis, called the dispersive model, suggested that DNA only copied itself for short chunks at a time, producing new strands that alternated parent and daughter DNA. Another idea, called the conservative model, argued that DNA didn't split open at all, but somehow kept the parent strands intact while creating an entirely new and separate copy.

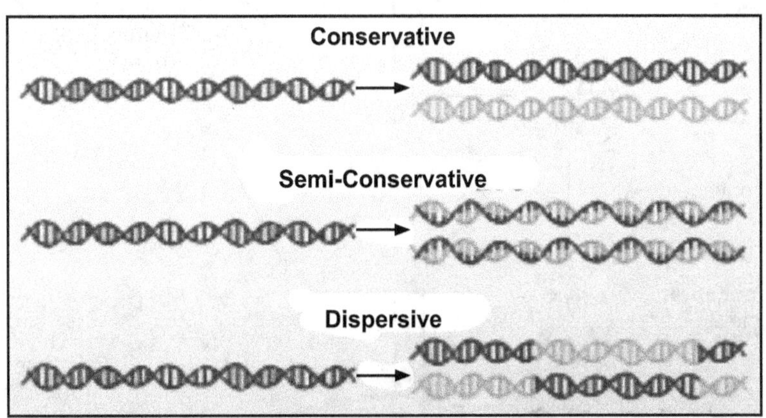

Fig. 16.1: The three proposed models of DNA replication

Nobody knew for sure how DNA replication really worked until two scientists named Matthew Meselson and Franklin Stahl devised an ingenious experiment in 1958. They realized that they could test all three models at once by keeping track of what happens to one parental DNA strand as it generates a series of copies.

Each model predicts a different distribution of parent DNA following a round of DNA replication. If Meselson and Stahl were able to keep track of parent versus new DNA, they could either support or refuse the predictions of the three different models.

The Meselson-Stahl Experiment:

Meselson and Stahl decided the best way to tag the parent DNA would be to change one of the atoms in the parent DNA molecule.

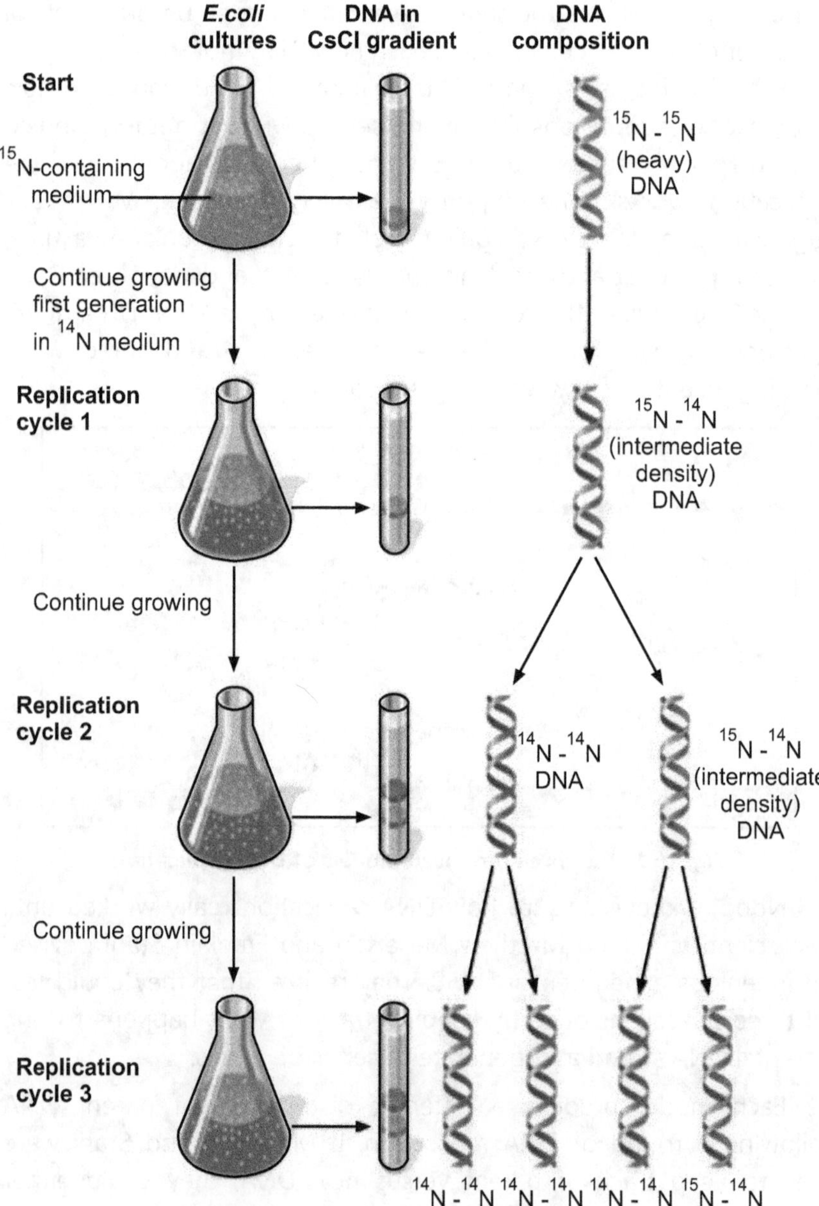

Fig. 16.2: Meselson-Stahl Experiment

Remember that nitrogen is found in the nitrogenous bases of each nucleotide. So they decided to use a heavy isotope of nitrogen (15N) to distinguish between parent and newly-copied DNA. The isotope of nitrogen had an extra neutron in the nucleus, which made it heavier.

You can see from any periodic table that most nitrogen atoms have an atomic weight of 14. We call these atoms N-14. But an isotope with an extra neutron has a weight of 15, so we call it N-15. The scientists decided to start with parent DNA molecules that only contained N-15. If only N-14 nucleotides were available during DNA replication, they would be able to tell which parts had come from the original double strand and which parts had been created during the replication process.

In order to make DNA go through many rounds of replication, Meselson and Stahl harnessed the reproductive powers of the common bacteria *E. coli.* They made sure that the first batch of bacteria contained only N-15 DNA. Then, they put the bacteria into a medium that only contained N-14 atoms. That way, whenever the bacteria reproduced, they would be forced to incorporate the N-14 into their new DNA. The scientists sat back and let the bacteria go to work.

With each new generation of bacteria, Meselson and Stahl took a sample so that they could see how the N-15 DNA was being distributed in the daughter molecules. Now, you may be wondering, how could they tell the difference between N-15 and N-14 DNA? It is not like you can actually see an isotope of nitrogen. How did the scientists know how much N-15 was inside each molecule?

The answer is the atomic weight. Because N-15 has one extra neutron, it is slightly heavier than N-14 and therefore makes the DNA molecule denser. We can separate DNA molecules based on the differences in their densities. To do this, we use a centrifuge, a device that spins a test tube at very high speeds. When a test tube is spun inside a centrifuge, all the contents are pushed toward the bottom. The substances that are heaviest sink farther down the tube, and the lighter substances float. So if you apply a centrifugal force to a mixture of two types of DNA, the heavier N-15 DNA sinks to a lower level than the N-14 molecules.

Every time Meselson and Stahl wanted to take a sample of the bacterial DNA, they had to chop up the tiny organisms and empty all the contents into a test tube. They mixed in a salt solution and then spun the test tube for many hours to make the substances separate out. Then they used special techniques to see how far the DNA molecules sank inside the tube.

When they sampled their first group of bacteria, Meselson and Stahl saw a darkened band in the test tube where the N-15 DNA had sunk and gathered in one spot. But after they let the bacteria reproduce, they got much different results in their samples. The DNA still sank down in the tube, but not nearly as far as the first generation. It was a lighter form of DNA, meaning that it was not completely made with the N-15 isotope. After one replication, the entire DNA had been converted to a hybrid of N-15 and N-14 DNA.

DNA must be replicated so that the information it holds can be maintained and passed to future cell generations, even as that information is accessed to guide the manufacture of proteins.

(a) Semiconservative model: All of the atoms of one strand of the parent molecule are transferred intact and without rearrangement to one strand of the progeny IDNA molecule: the other strand is formed entirely of new atoms.

(b) Conservative model: Atoms of the two parental DNA strands serve as a template for two new progeny strands. The two parental strands remain intact (without rearrangement) and remain together following replication, as do the two progeny strands.

(c) Dispersive model: All of the atoms of each strand of the parent molecule appear in the progeny DNA, but they appear as large sections scattered throughout the length of both strands of the progeny DNA.

Prokaryotic DNA replication:

Following are the components required for Prokaryotic DNA Replication:

1. Template DNA: The parent sequence of nucleotides to be used as information in the synthesis of the complementary strand. No any DNA polymerase enzyme is capable of synthesizing DNA de novo and nucleotides are never placed in a random order. A template is necessary, as the new DNA strand is being synthesized;

each new nucleotide to be added is matching to the nucleotide on the template strand based on Chargaff's base pairing rule. It is then, the sequence of nucleotides of the parent DNA strand that determines the sequence of nucleotides of the new DNA strand.

2. Origin of replication: Initiation of DNA synthesis does not begin just anywhere along the molecule. Rather, replication begins at specific locations or sites referred to as replication origins. Prokaryotes, has a single origin site, referred to as (oriC). At this site, initiation of replication occurs.

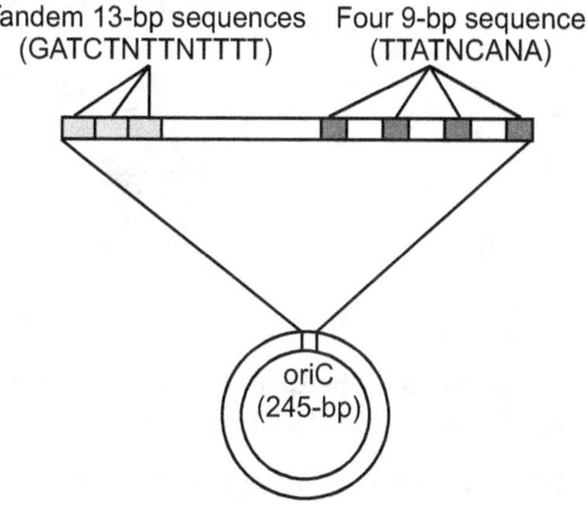

Fig. 16.3: Origin of replication of *E. coli*

Eukaryotic cells employ multiple sites on each chromosome that act as origins for the initiation of DNA replication. These sites are referred to as origins of DNA replication (ORIs).

3. Opening of DNA at the OriC site: DnaA, DnaB, and DnaC: Needed to recognize the origin and to separate the strands of the parent DNA. Helicase enzyme now binds to single stranded DNA and helps unwinding double stranded DNA throughout the process of replication. Single-strand binding proteins (SSB) also bind to the open single stranded DNA to avoid rewinding of parent DNA.

4. Nucleotides: The synthesis of DNA requires four nucleotides in the deoxyribose triphosphate form and four in the ribose triphosphate form. Nucleotides needed for DNA synthesis are; 1. deoxyadenine 5'-triphosphate (dATP) 2. deoxyguanosine 5'-triphosphate (dGTP) 3. deoxycytidine 5' triphosphate (dCTP) 4. deoxythymine 5'-triphosphate (dTTP).

Nucleotides needed for the synthesis of the RNAprimerare: 1. adenosine 5'-triphosphate (ATP) 2. guanosine 5'-triphosphate (GTP) 3. cytidine 5'-triphosphate (CTP) 4. uridine 5-triphosphate (UTP).

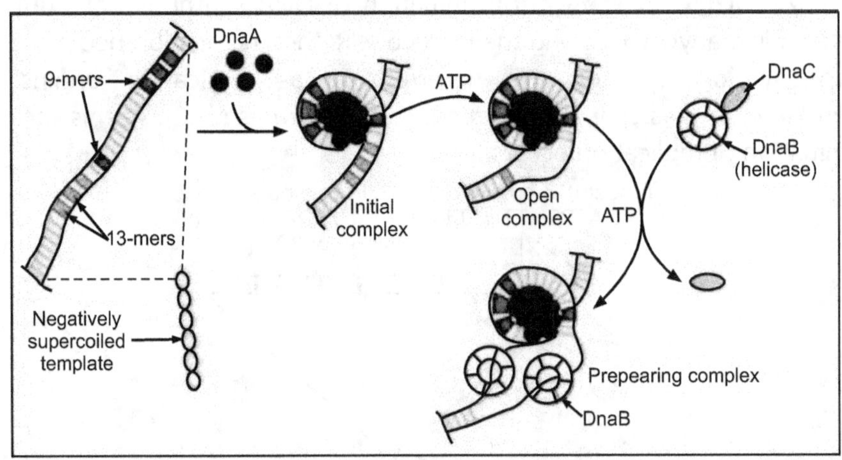

Fig. 16.4: Unwinding of the DNA at OriC site by DnaA proteins and binding of DnaB (i.e. Helicase) proteins to the unwound DNA.

5. Enzymes: Many enzymes are needed for replication, but we will deal only with the most important ones and in the order they are used: Gyrases uncoil the DNA in preparation for the activity of the helicase(s), helicases: Unwind the double helix. A site where DNA is locally opened, resembling a fork, is called a replication fork. RNA polymerase and primase: Needed for the synthesis of RNA primers, a requirement of DNA synthesis. DNA synthesis is initiated by synthesis. A short piece of RNA, called primer RNA because it primes DNA synthesis, is needed before DNA chain formation can begin. The primer varies in length in different cell types, but on the order of 2-10 nucleotides. These enzymes initiate synthesis and elongate chain only in 5' to 3' direction. DNA polymerase III: The actual replicating enzyme that synthesizes entire DNA.

DNA polymerase III is one of the two actual replicating enzymes. As its name implies it takes its instructions from the parental-template DNA strand. Using the RNA primer as an anchor and the triphosphate deoxyribonucleotides (dATTdGTP, dCTP, and dTTT), DNA polymerase adds deoxyribonucleotides to the primer, thereby elongating the chain. As the enzyme moves along the parent strand, it "reads" each parent nucleotide and adds new nucleotides to the growing DNA chain. Synthesis is always in the 5' to 3' direction.

DNA polymerase I: Removes the RNA primers and replaces them with DNA. It is also used in DNA repair systems. DNA polymerase I has, in fact, three enzyme activities:

1. DNA polymerization
2. Exonuclease activity in a 5' to 3' direction, used for the removal of RNA primer.
3. Exonuclease activity in a 3' to 5' direction, the proofreading activity.

DNA ligase: Joins the Okazaki fragments by making phosphodiester bond in between two DNA fragments.

The DNA ligases are responsible for connecting DNA segments during replication, repair, and recombination. All ligases join a 5'-phosphoryl group and a 3'-phosphoryl group on adjacent fragments, thereby sealing the nick. The reaction occurs in discrete steps and requires an energy source, either ATP (eukaryotes) or NAD (prokaryotes).

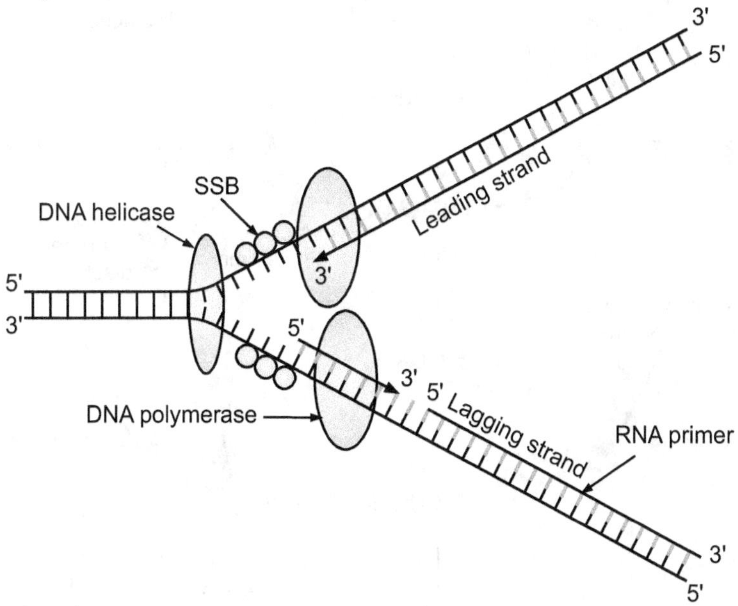

Fig. 16.5: Continuous and Discontinuous Synthesis of DNA strands

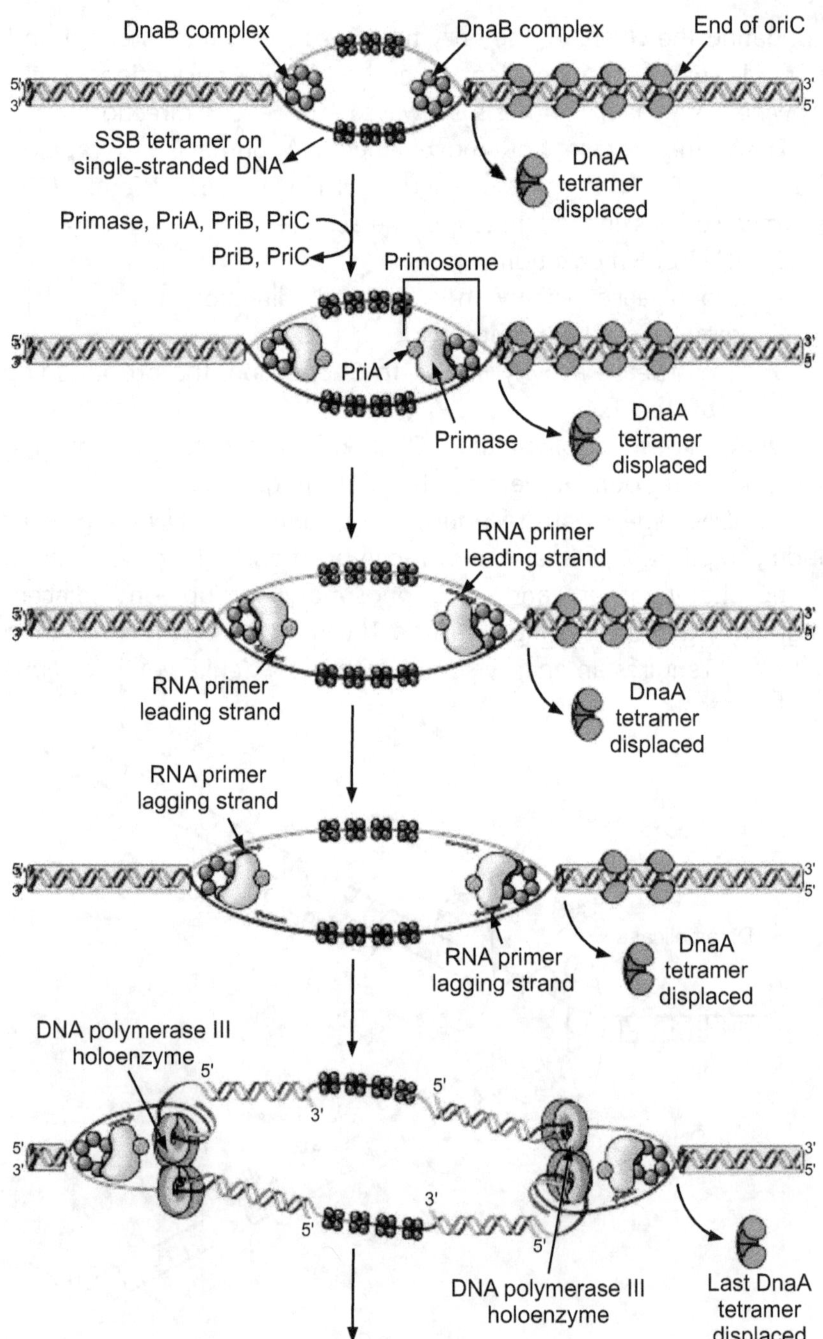

Fig. 16.6: Steps involved in Prokaryotic DNA replication

DNA synthesis process has to do with the way in which each new strand is synthesized. In this process, one strand is elongated by the continuous additions of nucleotides to the 3' end, this newly synthesized DNA is called the leading strand while the other new strand is produced by repeated synthesis of primer RNAs and short lengths of DNA (Okazaki fragments), which must eventually be joined by DNA ligase. This latter method is called discontinuous synthesis of the lagging strand. Since the DNA synthesis is always in the 5' to 3' direction and also the two strands of DNA are antiparallel; they run in opposite directions, the process of DNA replication is discontinuous on lagging strand. Okazaki fragments are short sequences of DNA nucleotides which are synthesized discontinuously and later linked together by the enzyme DNA ligase to create the lagging strand during DNA replication.

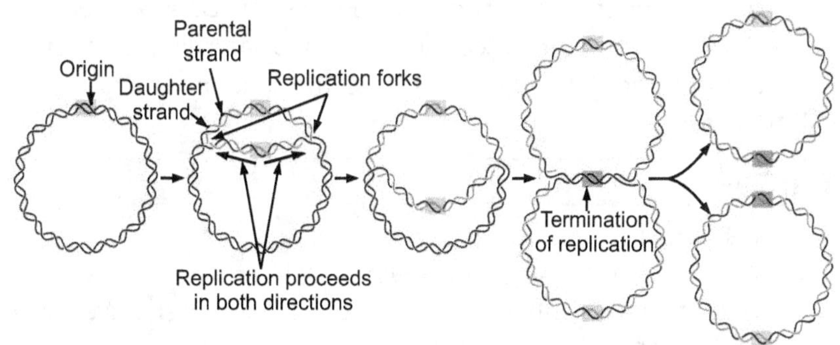

Fig. 16.7: Overall process of Prokaryotic DNA replication

POINTS TO REMEMBER

- The DNA replication is copying of sequence of parent DNA strand into daughter strand. In conservative DNA replication.

- DNA didn't split open at all, but somehow kept the parent strands intact while creating an entirely new and separate copy. In semi-conservative DNA replication half of the parent DNA is conserved in each new DNA molecule. During dispersive DNA replication.

- DNA only copied itself for short chunks at a time, producing new strands that alternated parent and daughter DNA.

- The components required for Prokaryotic DNA Replication includes:
  - Template DNA,
  - Replication origins,
  - Opening of DNA at the OriC site,
  - Nucleotides,
  - Enzymes.
- The parent strand of DNA which runs in the 3' to 5' direction toward the fork, and it is able to be replicated continuously by DNA polymerase is called leading strand.
- The parentstrand of DNA which runs in the 5' to 3' opposite to the direction of the growing replication fork, and it's able to be replicated discontinuously is called lagging strand.
- Okazaki fragments are short sequences of DNA nucleotides which are synthesized discontinuously and later linked together by the enzyme DNA ligase to create the lagging strand during DNA replication.

EXERCISE

1. Describe the process of DNA replication.
2. Explain the process of conservative DNA replication.
3. Give an account of process of semi-conservative DNA replication.
4. Write briefly the process of dispersive DNA replication.
5. Comment on the process of bacterial DNA replication.
6. Write short notes on:
   (a) Leading strand
   (b) Lagging strand
   (c) Okazaki fragments
7. Describe briefly the over cell process of prokaryotic DNA replication.
8. Comment on DNA polymerase and DNA ligase.
9. Give an account of enzymes in playing role in DNA replication.
10. Explain: Templet DNA.

❖❖❖

References

1. Buchanan, B.B, Gruissem, W. and Jones, R.L (2000). Biochemistry and Molecular Biology of Plants. American Society of Plant Physiologists Maryland, USA.
2. Cooper, G.M. and Hausman, R.E. (2007). The Cell: Molecular Approach 4th Edition, Sinauer Associates, USA.
3. David, Nelson and Cox, Michael (2007). Lehninger Principles of Biochemistry. W.H. Freeman and Company. New York.
4. Devlin, R.M. (1983). Fundamentals of Plant Physiology. Mc. Millan, New York.
5. Dutta, A.C. (2000). A Class Book of Botany. Oxford University Press, UK.
6. Hopkins, William G. (1995). Introduction to Plant Physiology. Publ. John Wiley and Sons, Inc.
7. Lewin, Benjamin (2011). Genes. X Jones and Bartlett.
8. Lincolin, Taiz and Eduardo, Zeiger (2010). Plant Physiology. 5th Edition. Sinauer Associates, Inc. Publishers. Sunder land, USA.
9. Opik, Helgi, Rolfe, Stephen A. and Willis, Arthur J. (2005). The Physiology of Flowering Plants. Cambridge University Press, UK.
10. Pal, J.K. and Ghaskadbi, Saroj (2009). Fundamentals of Molecular Biology. Oxford University Press. India.
11. Pandey, S.N. and Sinha, B.K. (2014). Plant Physiology. Vikas Publishing House Pvt. Ltd., India.
12. Salisbury, F.B. and Ross, C.B. (2005). Plant Physiology. 5th Edition. Wadsworth Publishing Co. Belmont California, USA.
13. Watson, James D., Baker, Tania; Bell, Stephen P.; Alexander Gann; Levine, Michael and Lodwick, Richard (2008). Molecular Biology of the Gene. 6th Edition, Pearson Education, Inc. and Dorling Kindersley Publishing, Inc. USA.
14. Weaver, R. (2011). Molecular Biology. 5th Edition, Publisher-McGrew Hill Science. USA.

PATTERN OF UNIVERSITY EXAM QUESTION PAPER

Total Marks : 35   **Duration : 2 Hours**

> **Note :** (a) Q. 1 is compulsory.
>
>   (b) Solve any three questions from Q.2 to Q.5.
>
>   (c) Questions 2 to 5 carry equal marks.

1.   **Attempt any five of the following :**

    (a)

    (b)

    (c)

    (d)

    (e)

    (f)

(Ask four tricky questions and two questions based on problem solving, if applicable.)   (5)

2.   (A) Describe – type of question(s) with internal options.   (6)

    (B) Short question, but tricky.   (4)

3.   (A) Explain – type of question(s) with internal options.   (6)

    (B) Problem based question, if applicable otherwise justification type – question.   (4)

4.   (A) Discuss - type of question(s) with internal options.   (6)

    (B) Problem based question, if applicable otherwise tricky and signified question.   (4)

5.   **Write short notes on any four of the following :**   (10)

    (A) Principle based.

    (B) System based.

    (C) Structure based.

    (D) Descriptive based.

    (E) Working based.

    (F) Model based.

❖❖❖

MODEL QUESTION PAPER-I

Total Marks - 35   *Duration – 2 hrs.*

NB:
(1) Question 1 is compulsory.
(2) Solve any Three questions from Q. 2 to Q. 5.
(3) Questions 2 to 5 carry equal marks.
(4) Draw well labelled diagrams wherever necessary.

Q.1 Answer Any Five of the following :   **(5)**
   (a) Give the definition of Osmosis.
   (b) Name the different phases of growth in plant.
   (c) What are the different phases of cell cycle?
   (d) What are the different model of Plasma Membrane?
   (e) Name the four nitrogen bases in DNA.
   (f) What are the different types of DNA replication?

Q.2 (A) Describe the Practical application of Auxins.   **(6)**

OR

Explain the concept of plasmolysis in plant cell. Add a note on its significance.

(B) Comment on the scope of plant physiology.   **(4)**

OR

Explain the different types of solutions.

Q.3 (A) Describe the stages of Mitosis in plant cell.   **(6)**

OR

Give detailed account of prophase I of meiosis.

(B) Explain the "Fluid mosaic model of plasma membrane sturcture".   **(4)**

OR

What are the functions of Mitochondria.

Q.4 (A) Describe the Watson-Crick Model of DNA structure.   **(6)**

OR

Explain the different types of RNAs.

(B) Describe the types of chromosomes based on the position of centromere.   **(4)**

OR

Explain the scope of molecular biology.

Q.5 Write short notes on Any Four of the following :   **(10)**

    (A) C value Paradox

    (B) Structure and function of t-RNA.

    (C) Factors affecting diffusion.

    (D) Applications of Gibberellins in Agriculture.

    (E) Sketch and label- Plant Cell.

    (F) Ultra Structure of chloroplast.

❖❖❖

MODEL QUESTION PAPER-II

Total Marks - 35   ***Duration – 2 hrs.***

Q.1 Answer Any Five of the following :   **(5)**

    (a) Define plant physiology.

    (b) What is Osmotic Pressure ?

    (c) Enlist the phases of growth.

    (d) Enlist the nitrogen bases of DNA/RNA.

    (e) Enlist types of chromosomes.

    (f) Enlist types of RNA.

Q.2 (A) Explain process of DNA replication.   **(6)**

    (B) Explain stages/phases of cell cycle.   **(4)**

Q.3 (A) Define plasmolysis. Write its mechanism and significance in plants.   **(6)**

    (B) Explain structures of tRNA.   **(4)**

Q.4 (A) Explain ultrastructure and functions of chloroplast/ mitochondria/endoplasmic reticulum.   **(6)**

    (B) Explain a mode of packaging of DNA into chromosomes. **(4)**

Q.5 Write short notes on Any Four of the following :   **(10)**

    (A) Stages of meiosis

    (B) Role of plant growth regulators

    (C) Osmosis

    (D) Characteristics of DNA

    (E) Transcription

    (F) Purine bases

❖❖❖

QUESTION BANK

CREDIT – I

1. Define plant physiology.
2. Define diffusion.
3. What is imbibition ?
4. Write significance of diffusion in plants.
5. Enlist factors affecting rate of diffusion.
6. Define Osmosis.
7. Enlist different types of solution.
8. What is Osmotic Pressure ?
9. Write a difference in turgor pressure and wall pressure.
10. What is difference in endoosomosis and exo-osmosis ?
11. Enlist the phases of growth.
12. What are different factors affecting growth ?
13. Give difference in prokaryotic and eukaryotic cell.
14. Name the components of primary and secondary cell wall.
15. What is the composition of plasma membrane ?
16. Write any two functions of chloroplast.
17. Write any two functions of mitochondria.
18. Write any two functions of endoplasmic reticulum.
19. Enlist the phases of cell cycle.
20. Give a difference in mitosis and meiosis.

Credit – II

1. Write a central dogma of molecular biology.
2. Give a scope of molecular biology.
3. Enlist the nitrogen bases of DNA/RNA.
4. What is nucleoside/nucleotide ?
5. State a law of Chargaff.
6. What is C-value paradox ?

7. Give characteristic features of B-form of DNA.

8. Give characteristic features of A-form of DNA.

9. Give characteristic features of Z-form of DNA.

10. Write a difference between B and Z form of DNA.

11. Enlist types of chromosomes.

12. What is nucleosome ?

13. What is the role of histone proteins ?

14. Enlist types of RNA.

15. What is role of tRNA/mRNA/rRNA ?

16. What are different types of replication of DNA ?

17. Enlist the names of enzymes involved in prokaryotic DNA replication.

18. What is Diffusion ? Give its importance in plants and factors affecting rate of diffusion.　　**6M**

19. What is osmosis ? Give its types and role of osmosis in plants.　　**6M**

20. Define plasmolysis. Write its mechanism and significance in plants.　　**6M**

21. What are plant growth regulators ? Write their role in plants.　　**4M/5M**

22. Write a fluid mosaic model of plasma membrane.　　**4M/5M**

23. Explain ultrastructure and functions of chloroplast/mitochondria/endoplasmic reticulum.　　**6M**

24. Explain stages of mitosis.　　**4M/5M**

25. Explain stages of meiosis.

26. Explain stages/phases of cell cycle.　　**4M/5M**

27. Write features of Watson and Crick model of DNA.　　**6M**

28. Explain types of Chromosomes.　　**4M/5M**

29. Explain structures of tRNA/mRNA.　　**4M/5M**

30. Explain a mode of packaging of DNA into chromosomes.　　**6M**

31. Explain process of DNA replication.　　**6M**

❖❖❖